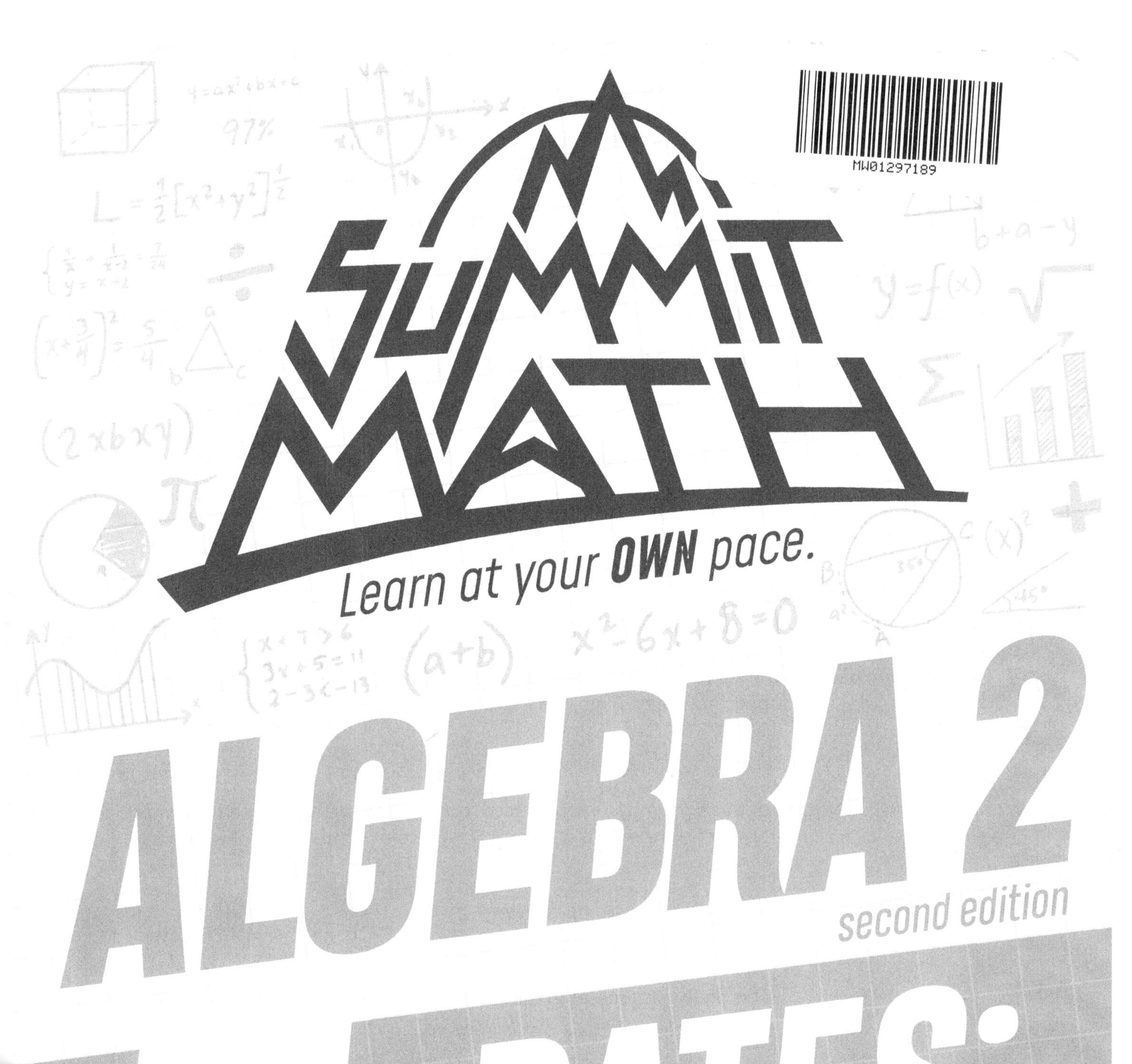

ALGEBRA 2

second edition

5 RATES: MOTION, WORK, & INTEREST

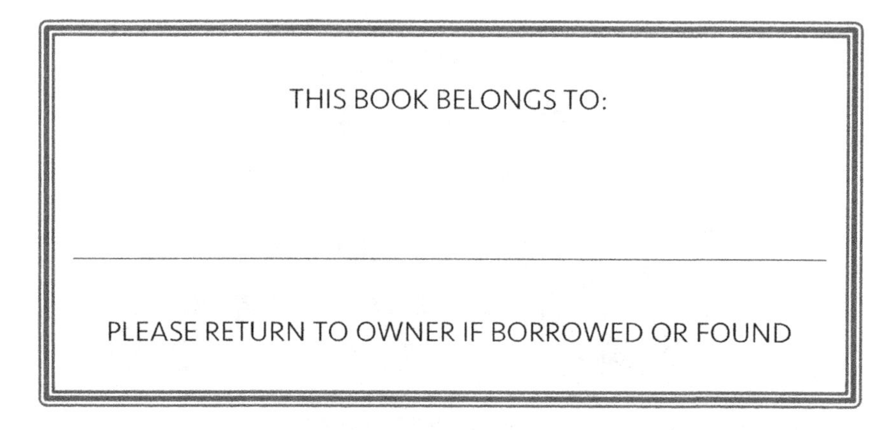

DEDICATION

To Lauren, Chloe, Dawson and Teagan

ACKNOWLEDGEMENTS

I started writing these books in 2013 to help my students learn better. I kept writing them because I received encouraging feedback from students, parents and teachers. Thank you to all who have used these books, pointed out my mistakes, and made suggestions along the way. Thank you to all of the students and parents who asked me to keep writing more books. Thank you to my family for supporting me through every step of this journey.

All rights reserved. No part of this book may be reproduced, transmitted, or stored in an information retrieval system in any form or by any means without prior written permission of the author.

Copyright © 2020

This book was typeset in the following fonts:
Seravek + Mohave + *Heading Pro*

Graphics in Summit Math books are made using the following resources:
Microsoft Excel | Microsoft Word | Desmos | Geogebra | Adobe Illustrator

First printed in 2017

Printed in the U.S.A.

Summit Math Books are written by Alex Joujan.

www.summitmathbooks.com

INTRODUCTION

Learning math through Guided Discovery:

A Guided Discovery learning experience is designed to help you experience a feeling of discovery as you learn each new topic.

Why this curriculum series is named Summit Math:

Learning through Guided Discovery can be compared to climbing a mountain. Climbing and learning both require effort and persistence. In both activities, people naturally move at different paces, but they can reach the summit if they keep moving forward. Whether you race rapidly through these books or step slowly through each scenario, this curriculum is designed to keep advancing your learning until you reach the end of the book.

Guided Discovery Scenarios:

The Guided Discovery Scenarios in this book are written and arranged to show you that new math concepts are related to previous concepts you have already learned. Try to fully understand each scenario before moving on to the next one. To do this, try the scenario on your own first, check your answer when you finish, and then fix any mistakes, if needed. Making mistakes and struggling are essential parts of the learning process.

Homework and Extra Practice Scenarios:

After you complete the scenarios in each Guided Discovery section, you may think you know those topics well, but over time, you will forget what you have learned. Extra practice will help you develop better retention of each topic. Use the Homework and Extra Practice Scenarios to improve your understanding and to increase your ability to retain what you have learned.

The Answer Key:

The Answer Key is included to promote learning. When you finish a scenario, you can get immediate feedback. When the Answer Key is not enough to help you fully understand a scenario, you should try to get additional guidance from another student or a teacher.

Star symbols:

Scenarios marked with a star symbol ★ can be used to provide you with additional challenges. Star scenarios are like detours on a hiking trail. They take more time, but you may enjoy the experience. If you skip scenarios marked with a star, you will still learn the core concepts of the book.

To learn more about Summit Math and to see more resources:

Visit www.summitmathbooks.com.

© Alex Joujan, 2020

GUIDED DISCOVERY SCENARIOS

As you complete scenarios in this part of the book, follow the steps below.

Step 1: Try the scenario.
Read through the scenario on your own or with other classmates. Examine the information carefully. Try to use what you already know to complete the scenario. Be willing to struggle.

Step 2: Check the Answer Key.
When you look at the Answer Key, it will help you see if you fully understand the math concepts involved in that scenario. It may teach you something new. It may show you that you need guidance from someone else.

Step 3: Fix your mistakes, if needed.
If there is something in the scenario that you do not fully understand, do something to help you understand it better. Go back through your work and try to find and fix your errors. Mistakes provide an opportunity to learn. If you need extra guidance, get help from another student or a teacher.

After Step 3, go to the next scenario and repeat this 3-step cycle.

NEED EXTRA HELP?
watch videos online

Teaching videos for every scenario in the Guided Discovery section of this book are available at www.summitmathbooks.com/algebra-2-videos.

 © Alex Joujan, 2020

CONTENTS

 © Alex Joujan, 2020

Section 1

INTRODUCTION TO RATES SCENARIOS

 © Alex Joujan, 2020

The concept of a rate should not be new to you, but there are many types of scenarios that involve rates. This book will guide you through several topics that involve rates.

1. According to one estimate, a 4-year-old child will ask 109,500 questions in one year (365 days). What is the rate at which a 4-year-old child asks questions, measured in questions per day?

2. At the beginning of the month, a family had used a total of 450 gallons of water since the water company started measuring their daily usage. Thirty days later, their total usage had increased to 630 gallons. What was the family's average daily water usage, in gallons per day?

3. Christine is allowed to use 2 Gigabytes of data every month. If she has used 1.4 GB of data at the end of the day on April 18th, will Christine go beyond her data usage limit for the month of April if she does not change her data usage rate?

4. Fred and Cathy both worked in their bakery to put icing on cupcakes for an upcoming event. Fred puts icing on F cupcakes every minute that he works and Cathy ices C cupcakes per minute. After they finished, the total number of cupcakes iced by Fred and Cathy was given by the expression $24F + 38C$.

    a. How many minutes did Fred and Cathy each spend putting icing on cupcakes, respectively?

    b. Who worked at a faster rate if Fred iced a total of 144 cupcakes and Cathy iced a total of 190 cupcakes?

 © Alex Joujan, 2020

5. A gas station has an underground storage tank that can hold 12,000 gallons of gas when it is completely full. At the end of a busy day, the tank only contains 4,000 gallons of gas. A gas truck arrives to refill the tank and when it arrives, it contains 9,000 gallons of gas. When the truck starts refilling the storage tank, the gas flows from the truck into the station's storage tank at a rate of 120 gallons per minute.

   a. How much gas is in the station's storage tank after it has been refilling for 12 minutes?

   b. How long does it take the truck to completely refill the station's storage tank?

   c. Write an expression that represents the amount of gas in the truck after the truck has been refilling the storage tank for M minutes.

6. Use your work in the previous scenario to answer the following questions.

   a. After how many minutes will the truck and the station's storage tank contain the same amount of gas?

   b. ★If the truck starts pumping gas into the storage tank at 8:00pm, at what time will the storage tank contain twice as much gas as the truck?

 © Alex Joujan, 2020

Use this page to record important ideas in the previous section or for any other writing that helps you learn the topics in this book.

 © Alex Joujan, 2020

Section 2

RATES AND THEIR RECIPROCALS

 © Alex Joujan, 2020

The next group of scenarios will lead you to compare a rate and the reciprocal of that rate in order to see that both of these forms of a rate can be useful.

7. If a rate is measured in miles per hour, it may sometimes be helpful to express the rate as a reciprocal.

   As a reciprocal, the units of this rate are ____________________________.

8. For example, the reciprocal of 10 miles per hour would be 1 hour per 10 miles. If you convert this to a unit rate, it becomes 0.1 hour per mile. Write the reciprocal of each rate shown below. Convert the reciprocal rate into a unit rate.

   a. 8 people per pizza

   b. 0.25 inch per minute

9. A runner travels a distance of 3.6 miles in 24 minutes.

   a. You can calculate that the runner moved at a rate of _____ miles per minute.

   b. However, it might be more useful to state that the runner maintained a running pace of _____ minutes per mile, or _____ minutes and _____ seconds per mile. Distance runners are often interested in the time that it takes to travel each mile over the course of a long run.

   c. What was the runner's speed, measured in miles per hour?

10. Ajay lives 24 miles from school and his car can travel 400 miles on 12 gallons of gas. How many gallons of gas will his car consume when he drives to school and back home again?

11. Consider the graph shown.

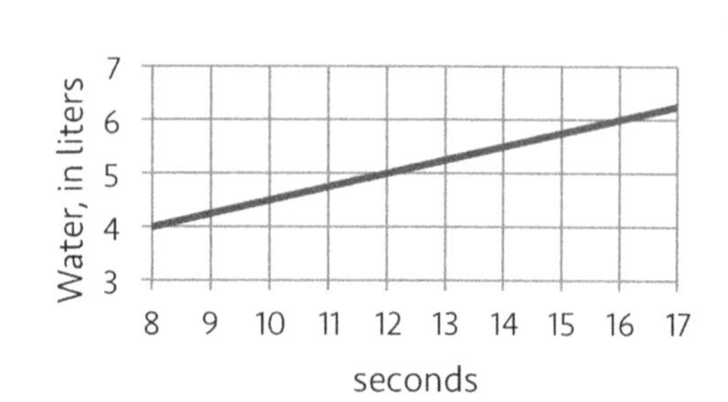

   a. Describe the rate shown by the data in the graph.

   b. Express the reciprocal of this rate, using proper units.

 © Alex Joujan, 2020

Use this page to record important ideas in the previous section or
for any other writing that helps you learn the topics in this book.

 © Alex Joujan, 2020

Section 3

SCENARIOS INVOLVING WORK RATES AND COMBINED WORK

 © Alex Joujan, 2020

12. In an effort to help out at a local fundraiser, you offer to wrap gifts for $2 each. You can wrap 1 gift in 4 minutes. What portion of a gift can you wrap every minute? (This is your wrapping rate, measured in gifts per minute.)

13. In the previous scenario, how much money would you earn for the fundraiser in 1 hour?

14. When it snows outside, you like to clear the snow off of your neighbor's driveway. It takes you 40 minutes to clear all of the snow.

    a. What is the rate at which you work, measured in driveways per minute?

    b. How much of the driveway do you clear off in 10 minutes?

    c. How much of the driveway do you clear off in M minutes?

15. Suppose it takes a pump 6 minutes to fill a gas tank.

    a. How much of the tank is filled after 1 minute?

    b. How much of the tank is filled after M minutes?

    c. What is the rate at which the pump fills the tank, measured in tanks per minute?

 © Alex Joujan, 2020

16. Caleb plants 60 blueberry bushes in 8 hours. Gerry plants 50 blueberry bushes in 4 hours.

a. What is Caleb's work rate?

b. What is Gerry's work rate?

c. When Gerry and Caleb work together, how many bushes can they plant in 1 hour?

d. If they work together, how long will it take them to plant 240 blueberry bushes?

17. If you can paint a room in 2 hours and your friend can also paint the entire room in 2 hours, estimate how long it will take to paint the room if you work together at the same time.

18. Read the scenario below and make a quick guess. Do not try to set up and solve an equation. The goal here is to see if you can get close just by guessing.

One caterpillar can eat an entire leaf in 12 hours. Another caterpillar can eat a leaf in 8 hours. Estimate the amount of time it would take for both caterpillars to eat the same leaf if they start at opposite ends.

19. Guess the missing information in the scenario below. Do not set up and solve an equation.

One beaver can build a dam in 4 days. Another beaver can build a dam in 6 days. If they work together at the same time, they can build a dam in _______ days.

 © Alex Joujan, 2020

20. Solve each equation below.

a. $\frac{1}{4}t+\frac{1}{3}t=1$

b. $\frac{1}{5}h+\frac{1}{8}h=1$

21. Whenever the grass gets too long, Harriet and Corwin take turns mowing the grass. If Corwin does all of the mowing by himself, it takes him 40 minutes to finish the job. When they use two mowers and work together at the same time, their work is modeled by the equation below.

$$\frac{1}{40}t+\frac{1}{30}t=1$$

In this equation, t is the amount of time that it takes them to mow the grass when they use two mowers and work together.

a. If Harriet does all of the mowing by herself, how long does it take her to finish the job?

b. How long does it take for Harriet and Corwin to do the mowing when they work together?

22. At the end of a football game, a lot of trash is left behind in the seating areas. Jim can clean up all of the trash in 5 hours. Becky can clean up all of the trash in 4 hours. How long will it take Jim and Becky to clean up all of the trash if they work together?

 © Alex Joujan, 2020

23. Two friends, Carl and Ryan, work together to wash each car that drives through their car wash. Carl can wash a car in 12 minutes, while Ryan can wash a car in 10 minutes.

a. How much can Carl wash in 1 minute? How much can Ryan wash in 1 minute?

b. When they work together, how much can Carl and Ryan wash in 2 minutes?

c. How much can Carl and Ryan wash in 3 minutes?

d. How much can they wash in m minutes?

e. How many minutes does it take them to wash one car?

24. ★In the previous scenario, how many <u>hours</u> does it take them to wash two cars?

25. A young girl in Zambia carries a bucket of water from her house to the river and back again in 9 minutes. Her older brother completes this round trip in 6 minutes.

a. Working together, how many buckets of water can they bring home in 1 hour?

b. Suppose their family needs 20 buckets of water. If the older brother works for 30 minutes by himself and then his sister joins him, how much time will they need to spend working together to collect the total amount of water that their family needs?

 © Alex Joujan, 2020

26. Students on a service trip to Haiti form two lines on a wide ramp and pass cinder blocks from a truck to the second floor of a school they are building. Anya leads one line, while Barry leads the other line. On Monday, Anya's line found that they could move a truckload of blocks in 30 minutes. When they work together with Barry's line on Tuesday, the two lines move a truckload of blocks in 12 minutes. How long should it take Barry's line to move a truckload of blocks if Barry's line works alone on Wednesday?

   a. First, guess the answer and write down your estimate.

   b. Write and solve an equation to answer this question.

27. Bette can build $\frac{2}{7}$ of a bicycle every hour. Sara can build $\frac{1}{5}$ of a bicycle every hour. If they both work for 40 hours one week, who builds the most bicycles and how many more does she build?

28. ★Ansley is trying to organize a fundraiser and she needs to send out a huge stack of invitations. She asks her friend to help her place the invitations in envelopes. She works twice as fast as her friend, and as a result, they are able to finish the entire job in 3 hours. How long would it have taken Ansley to finish the job if she was working alone?

 © Alex Joujan, 2020

Use this page to record important ideas in the previous section or
for any other writing that helps you learn the topics in this book.

© Alex Joujan, 2020

Section 4

SCENARIOS INVOLVING MOTION

 © Alex Joujan, 2020

29. For your birthday, someone gave you a cyclometer, which is a device that shows your speed and total distance as you ride. One morning, you start a long ride at 8:00am and don't get back home until 9:30am. Looking through your cyclometer data, you see that you maintained an average speed of 18 miles per hour. How many miles did you cover on this bike ride?

30. If you run one lap around a 400-meter track, how can you determine how fast you ran?

31. If a pilot checks her flight statistics after a 600 mile flight and finds that her average air speed was 240 miles per hour, how can she determine the total flying time of that trip, in hours?

32. If you go on a long road trip with your family for T hours and the car averages a speed of R miles per hour, what expression represents the total distance that you travel on this road trip?

33. A young water bug drops from a branch and lands in a lake, 10 feet in front of a hungry fish swimming along at 2 feet per second. If the fish does not see the water bug and the water bug does not move, how long will it take the fish to reach the water bug?

 © Alex Joujan, 2020

34. ★The previous scenario is unlikely, though, because as soon as the water bug hits the water, it will know its life is in danger and it will start skimming the top of the water to try to get to safety. If the water bug hits the water and immediately starts moving away from the fish in the same direction as the fish is swimming (bad choice, little guy), how fast is the water bug skimming if the fish catches the water bug in 8 seconds?

35. The previous scenarios are all designed to help you see that the distance traveled by a person or object is the product of the speed and the time spent traveling. In a very concise form, this relationship can be stated as $D = RT$. Though the equation $D = RT$ shows the relationship between distance, rate, and time, this equation can also be written in other forms.

    a. Restate the relationship to isolate R.

    b. Restate the relationship again to isolate T.

 © Alex Joujan, 2020

36. Two jousting knights, Sir Morton and Sir Lester, line up on opposite ends of a field to engage in a duel. The length of the field is 500 meters. Sir Morton's horse rides at an average pace of 20 meters per second, while Sir Lester's horse rides at an average pace of 12 meters per second. When the official signals for the knights to begin riding, Sir Morton gets tangled up in his armor and leaves 5 seconds later than Sir Lester.

Using what you have learned about distance, rate, and time, you can figure out how long it takes for the knights to collide.

a. What expressions represent the separate times, in seconds, that each knight spends in motion on his horse?

b. What expressions represent the distances each knight travels, in meters?

c. Use your expressions in part b. to create an equation that relates all of the distances in this scenario. Solve this equation to figure out how many seconds it takes for Sir Morton to strike Sir Lester.

37. In the previous scenario, use the relationship between distance, rate, and time to determine which knight rides farther.

 © Alex Joujan, 2020

38. On Saturday morning, you decide to bike to your friend's house 23 miles away. It turns out that your friend decides to bike to your house as well. Neither of you tells the other you are coming. You leave at 9:00am and maintain an average speed of 15 miles per hour. Your friend leaves 15 minutes later and maintains an average speed of 18 miles per hour.

Using what you have learned about distance, rate, and time, you can figure out how long it takes for you to pass your friend on your bike.

a. What expressions represent the separate times, in hours, that you each spend riding your bikes?

b. What expressions represent the distances you each travel?

c. Use your expressions in part b. to create an equation that relates all of the distances in this scenario. Solve this equation to figure out what time it is when you pass each other.

39. Use the relationship between distance, rate, and time to determine how far you rode before you passed your friend in the previous scenario.

 © Alex Joujan, 2020

40. During Summer Break, your family decides to drive to the shore to spend some time at the beach. You have been eager to go on a long bike ride and realize that this trip gives you that chance. You fill up a couple of water bottles and head out on your bike at 7:00am, while your family leaves around 9:00am and follows roughly the same route. You maintain an average speed of 15 miles per hour, while your family drives at an average speed of 55 miles per hour.

    a. If you and your family arrive at the beach at the same time, what time did you arrive?

    b. How many miles did you ride on your bike?

41. ★To make a delivery, a driver traveled from her workplace to a client's business at 48 miles per hour and returned along the same route at 24 miles per hour, due to traffic. The entire trip took 4 hours.

    a. If the driver spent h hours driving to the client's business, what expression represents the time it took to drive back to her workplace? Check the validity of your expression by confirming that your two time expressions add up to 4 hours.

    b. What expressions represent the distances the driver travels in each direction?

    c. Create an equation that relates the two distances in this scenario. Solve the equation to determine how many hours it took for the driver to travel to the client's business.

 © Alex Joujan, 2020

42. In the previous scenario, how many miles did the driver travel for the entire round trip journey?

43. Dan sits in his car waiting for a train to pass. He starts counting the individual boxcars of the train as they pass in front of him. In one minute, Dan counts 30 boxcars. This makes Dan curious. He does some research later and finds out that a train boxcar is 60 feet long. How fast was the train moving?

 © Alex Joujan, 2020

Use this page to record important ideas in the previous section or for any other writing that helps you learn the topics in this book.

 © Alex Joujan, 2020

Section 5
CALCULATING AVERAGE SPEED

 © Alex Joujan, 2020

44. If you ride in a plane for 3 hours and the flight takes you a distance of 1800 miles, you have enough information to calculate the average speed of the plane during this trip. Your average speed would be ______ miles per hour.

Since the plane speeds up at the start and slows down at the end, it does not fly at one speed for the entire trip, but you can still calculate the average speed to get an estimate of the speed of the plane.

45. A migrating flock of geese can fly 1000 miles in 25 hours. Although the speed of the geese may change as they fly, what is the average flying speed of the geese during their migration?

46. What information do you need in order to calculate the average speed of an object in motion?

47. Diana runs around town for 2.5 hours and she maintains a speed of 8 miles per hour. She then walks slowly for one half of an hour, moving at a rate of 2 miles per hour.

a. A quick mental calculation reveals that the average of 8 miles per hour and 2 miles per hour is 5 miles per hour. Why is it unreasonable to state that Diana's average speed for the 3 hours that she was moving was 5 miles per hour?

b. What was the total distance that Diana traveled during her 3 hours of exercise?

c. What was Diana's average speed over the course of her 3 hours of motion?

 © Alex Joujan, 2020

48. Zane drove from his apartment in the city to his grandmother's house out in the country. Driving through the heavy traffic of the city, Zane maintained an average speed of 30 km/h for 2 hours. When he reached the open country roads, he increased his average speed to 50 km/h for the next 3 hours until he arrived at her house. What was his average speed for the entire trip?

49. Harold took a trip in his private plane. He flew 1000 miles in 5 hours. He maintained an average speed of 240 miles per hour for the first 3 hours of the flight, but he slowed down after that because he was flying in a storm. What was Harold's average speed for the rest of the flight?

50. A stunt diver stands at the top of a waterfall and then jumps upward and outward. As she falls, her dive is modeled by the graph shown.

a. Estimate the initial height of the diver before she jumped.

b. What was the average falling speed of the diver, in feet per second, during the final half-second before she hit the surface of the water?

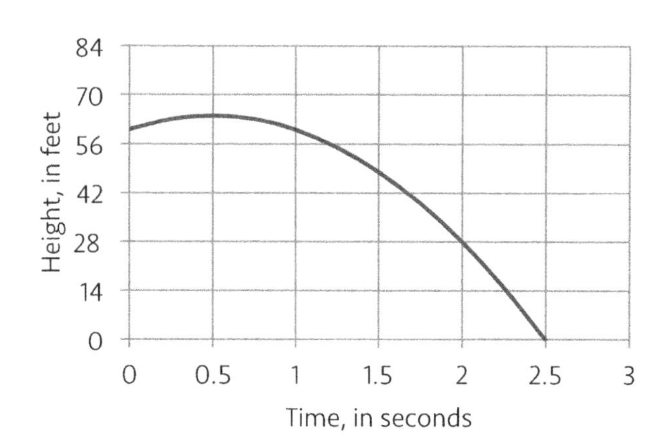

c. How long was the diver falling before she hit the surface of the water?

51. In the previous scenario, why is it unreasonable to discuss the average rate at which the diver falls over the course of her entire jump?

 © Alex Joujan, 2020

Use this page to record important ideas in the previous section or for any other writing that helps you learn the topics in this book.

© Alex Joujan, 2020

Section 6

SCENARIOS INVOLVING INTEREST RATES

 © Alex Joujan, 2020

52. If you put your money in an account that earns interest, the value of your money increases over time. For example, suppose you put $1,000 in a savings account that has a 2% interest rate. If you leave your money in the account for one year, its value will increase by 2%.

    a. After one year, your account is now worth $________.

    b. The value of your account increases by $________ during the first year.

    c. How much does the value of your account increase during the second year?

53. If someone has $1,000 in a savings account and they add money at a rate of $50 per day, how much money will be in the account after 10 days of increases?

54. If someone has $1,000 in a savings account and they add money at a rate of 5% per day, how much money will be in the account after 10 days of increases?

55. Helen has $10,000 in her bank account. Sam needs to borrow money to buy a car. He asks Helen for $5,000. Helen gives Sam $5,000, but she wants to earn some money in the process so she makes him agree to pay her an extra $10 per month until he has paid back all that he owes.

    a. If it takes Sam two full years to pay back what he borrowed, how much money does Helen earn for loaning Sam $5,000?

    b. By what percent did Helen increase the value of her $5,000 loan by making Sam pay extra?

 © Alex Joujan, 2020

56. Helen now has $10,240 in her bank account, but the value of that money will not change over time. In the mail one day, she receives a letter from her bank that asks her if she would like to move her money into a savings account that will pay her to keep her money in that account. Helen has no other plans for this money so she transfers the money to a savings account. In return for giving the bank her money, the bank pays Helen by increasing her account value by 2% every year. To keep this simple, assume that the bank pays Helen only at the end of each year.

    a. How much money will Helen earn at the end of the first year?

    b. If Helen puts that money back into her savings account, how much will the bank pay Helen at the end of the second year?

57. Continue your work in the previous scenario.

    a. If Helen keeps putting the extra money back into her account, how much will the bank pay her at the end of the fourth year?

    b. What is the total value of Helen's savings account at the end of the fourth year?

58. Helen started with $10,240 in her savings account. After 4 years of 2% increases, she had $11,084.11 in her savings account. By what percent did her savings account increase over those 4 years? Do 4 years of 2% increases add up to make a total increase of 8%?

 © Alex Joujan, 2020

59. For her 21st birthday, Gail's generous uncle opens an investment account for her and deposits $500 in the account. Her generous aunt also opens an account for her and deposits $800 in the account. After one year, the uncle's account earns 16% in interest, while the aunt's account earns 10% in interest.

    a. By what percent did the total value of Gail's two accounts increase that year?

    b. If the amounts of money that were placed in the two accounts were switched, by what percent would the total value of Gail's accounts have increased that year?

60. Several years later, Gail has a total of $6,000 in her two accounts. During that year, she earns 6% interest on the money in one account and 9.4% interest on the money in her other account. If she earns a total of $445 in interest that year, how much money did she have in each of the accounts that year?

61. Gail's friend Haley saw how Gail's investments were increasing and she decided to open some accounts as well. She invests $4,000 in two different types of accounts. Haley invests some of the money in a mutual fund and then she uses the rest of the money to buy a government savings bond. Over the next year, the mutual fund grows by 8% and the value of the savings bond increases by 3.75%. At the end of the year, the total value of her investments is $4,269. How much did Haley invest in the savings bond?

 © Alex Joujan, 2020

62. Luc invests a total of $14,000. He put some of the money in a mutual fund and he puts the rest in bonds. The equations below show the relationship between the amount of money he invested and the amount of interest he earned after one year.

Equation 1: $m + b = 14{,}000$
Equation 2: $0.11m + 0.025b = 881.25$

a. What was the combined amount that Luc earned in interest after one year?

b. By what percent did the amount of money he invested in bonds increase after one year?

c. How much money did Luc invest in a mutual fund?

63. Your income from summer jobs over the years has allowed you to accumulate $15,000 in savings. You take that money and invest it in 3 different accounts. After one year, the 3 accounts earn 2%, 5%, and 8% in interest, respectively. As a result of these interest rates, the total value of your investments increases by $720 that year. You invested $4000 more dollars in the account that earned 5% than in the account that earned 2%.

Write three equations that represent the relationships described in this scenario.

 © Alex Joujan, 2020

Use this page to record important ideas in the previous section or for any other writing that helps you learn the topics in this book.

 © Alex Joujan, 2020

Section 7
CUMULATIVE REVIEW

 © Alex Joujan, 2020

64. In order to analyze how pricing affects a person's purchasing decisions, a company packages water in 3 different bottles and sells the water in an amusement park on a hot summer day. The water in a blue bottle sells for $2, while the green bottle is $3 and the yellow bottle is $3.50. At the end of the day, they had sold 19,200 bottles and earned $56,400 in revenue. The number of green bottles purchased was three times larger than the number of blue bottles purchased.

Write three equations that represent the relationships described in this scenario.

65. A business in Germany mails a total of 3,000 fundraising letters to 3 different countries, England, Poland, and France. They pay postage costs of $1.25 for each letter sent to England, $1.40 per letter sent to Poland, and $1.75 per letter sent to France. The number of letters sent to France was the same as the number of letters sent to England. The total postage costs were $4,420.

Write three equations that represent the relationships described in this scenario.

66. Solve the equation.

$$\frac{1}{x^2-2x-8}-\frac{1}{x+2}=\frac{7-3x}{x^2-2x-8}$$

67. How far apart are 13 and 30? How far apart are 14 and 40? What about 15 and 50, 16 and 60, and so on? Analyze the distances between these numbers and see if there is a pattern that you can identify.

 © Alex Joujan, 2020

68. Posts are placed as shown in the figure. Each circle represents a post. A fence is built by connecting the posts. Arrows are drawn to show length measurements. The figure is not drawn to scale. How long is the perimeter of the fence?

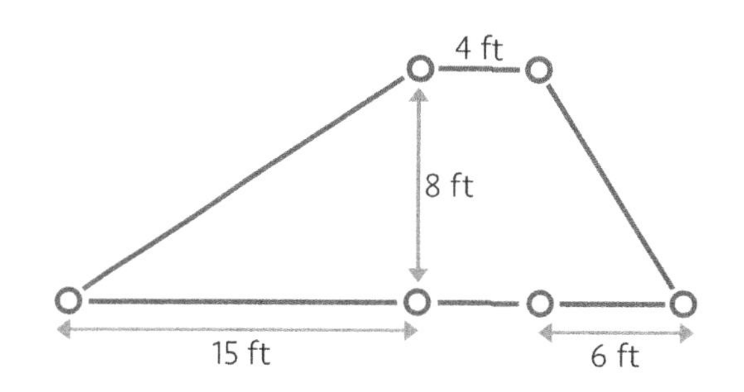

69. If the fence in the previous scenario is 6 feet high, how many square feet of paint will be needed to cover the entire outside surface of the fence if it is painted?

70. Calculate the area of the ground enclosed by the fence in the previous scenario.

71. A function is defined as shown below. The function is created by using parts of 3 separate functions. Graph the function.

$y=4 \quad \text{if } x<-4$

$y=-x \quad \text{if } -4\leq x\leq 0$

$y=x^2 \quad \text{if } x>0$

To graph this function, first graph all 3 equations on the same grid and then erase the parts of them that are not included. Use the inequalities to find the section of each equation that you should erase.

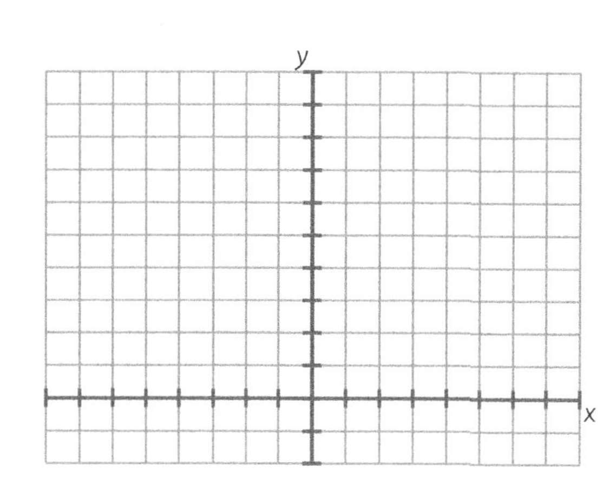

 © Alex Joujan, 2020

Use this page to record important ideas in the previous section or for any other writing that helps you learn the topics in this book.

 © Alex Joujan, 2020

Section 8
ANSWER KEY

1.	300 questions per day
2.	6 gallons/day (180 gal. over 30 days)
3.	Yes, she will use approx. $2\frac{1}{3}$ GB of data.
4.	a. Fred: 24 min; Cathy: 38 min b. Fred (6 per min.) > Cathy (5 per min.)
5.	a. 4000 + 120(12) = 5,440 gallons b. $66\frac{2}{3}$ minutes $\rightarrow$ 1 hr, 6 min, 40 sec c. 9,000 – 120*M*
6.	a. solve: 4000 + 120*M* = 9000 – 120*M*; $M = 20\frac{5}{6}$ minutes – 20 min, 50 sec b. solve: 4000 + 120*M* = 2(9000 – 120*M*); $M = 38\frac{8}{9}$ minutes – approx. 8:39pm
7.	hours per mile
8.	a. $\frac{1}{8}$ of a pizza per person b. 4 minutes per inch
9.	a. 0.15 mi./min b. $6\frac{2}{3}$ min/mi. or 6 min, 40 sec. per mile c. 9 mph
10.	1.44 gallons (48 miles at $33\frac{1}{3}$ mi per gal.)
11.	a. 0.25 liters of water per second b. 4 seconds per liter of water
12.	$\frac{1}{4}$ of a gift per minute
13.	1 gift every 4 minutes for is a total of 15 gifts in 1 hour, so you earn $30.
14.	a. $\frac{1}{40}$ driveway per minute b. $\frac{10}{40}=\frac{1}{4}$ c. $\frac{1}{40}M$ or $\frac{M}{40}$
15.	a. $\frac{1}{6}$ b. $\frac{m}{6}$ c. $\frac{1}{6}$ tank per minute
16.	a. 7.5 bushes per hr b. 12.5 bushes/hr c. 20 bushes d. Their combined work rate is 20 bushes per hour, so it will take them 12 hours.
17.	1 hour
18.	Your guess should be between 4 and 6 hours. The exact time is 4.8 hours...you'll learn how to find this later.
19.	Your guess should be between 2 and 3 days. The exact time is 2.4 days...you'll learn how to find this later.
20.	a. $\frac{3}{12}t+\frac{4}{12}t=1 \rightarrow \frac{7}{12}t=1 \rightarrow t=\frac{12}{7}$ or $1\frac{5}{7}$ b. $\frac{8}{40}h+\frac{5}{40}h=1 \rightarrow \frac{13}{40}h=1$ $\rightarrow h=\frac{40}{13}$ or $3\frac{1}{13}$
21.	a. 30 minutes b. $17\frac{1}{7}$ min solve $\frac{1}{40}t+\frac{1}{30}t=1$; $H=\frac{120}{7}$ $\left(17\frac{1}{7}\text{ min}\right)$
22.	solve $\frac{1}{4}H+\frac{1}{5}H=1$; $H=\frac{20}{9}$ $\left(2\frac{2}{9}\text{ hrs}\right)$
23.	a. Carl = $\frac{1}{12}$, Ryan = $\frac{1}{10}$ b. $\frac{11}{30}$ c. $\frac{11}{20}$ d. $\frac{11m}{60}$ e. $5\frac{5}{11}$ min (5 min, 27 sec)
24.	$\frac{2}{11}$ of a hour
25.	a. 16 b. 54 minutes
26.	a. Estimates close to 20 are accurate. b. $\frac{1}{30}(12)+\frac{1}{B}(12)=1 \rightarrow$ 20 minutes
27.	Bette builds 1 bike every 3.5 hours, or 11.4 per week. Sara builds 1 bike every 5 hours, or 8 per week. In 1 week, Bette builds about 3.4 more bicycles than Sara.
28.	Equation: $\frac{1}{x}(3)+\frac{2}{x}(3)=1$; 4.5 hours
29.	27 miles
30.	distance ÷ time
31.	600 ÷ 240

 © Alex Joujan, 2020

32.	distance = RT miles
33.	5 seconds
34.	9 inches per second
35.	a. $R=\frac{D}{T}$ b. $T=\frac{D}{R}$
36.	a. Sir M: $t-5$ Sir L: t b. Sir M: $d=20(t-5)$ Sir L: $d=12t$ c. solve: $20(t-5)+12t=500 \rightarrow t=18.75$sec Sir Morton rides for 13.75 seconds
37.	Sir Morton: 275m Sir Lester: 225m Sir Morton travels 50m farther
38.	a. $t,\ t-\frac{1}{4}$ b. $15t,\ 18\left(t-\frac{1}{4}\right)$ c. $15t+18\left(t-\frac{1}{4}\right)=23;\ t=\frac{5}{6}$ hr; 9:50am
39.	$15\cdot\frac{5}{6}\rightarrow 12.5$ miles
40.	a. 9:45 AM b. 41.25 miles
41.	a. $4-h$ b. $48h,\ 24(4-h)$ c. $48h=24(4-h)$ $h=1\frac{1}{3}$ hrs to get to the client's business
42.	$48\cdot\frac{4}{3}\rightarrow 64\rightarrow 64\cdot 2=128$ miles
43.	30 boxcars per minute = 1800 feet/min = 108,000 ft/hr $\approx$ 20.5 miles per hour
44.	600 miles per hour
45.	40 miles per hour
46.	Total distance traveled and total time spent moving
47.	a. She spent much more time traveling 8 miles per hour, so her average speed is not the average of her 2 separate speeds. b. 20 miles + 1 mile = 21 miles c. 21 miles over 3 hours is 7 mph.
48.	usual mistake: $(30+50)\div 2=40$kmph actual result: 210km$\div$5hr = 42kmph
49.	140 miles per hour He flew 280 miles in the last 2 hours.
50.	a. 60 ft b. 56 ft/s (28 ft in 0.5 sec.) c. 2 sec. (not falling for the first 0.5 sec.
51.	She is not falling the entire time.
52.	a. \$1,020 b. \$20 c. \$20.40
53.	\$1,500
54.	Approx. \$1,628.89
55.	a. \$240 b. 4.8%
56.	a. \$204.80 b. \$208.90
57.	a. \$217.34 b. \$11,084.11
58.	Approx. 8.24%
59.	a. about 12.3% b. about 13.7%

60.	\$3500 in the account earning 6% \$2500 in the account earning 9.4% Equation 1: $a+b=6000$ Equation 2: $0.06a+0.094b=445$
61.	\$1200 Equation 1: $a+b=4000$ Equation 2: $1.08a+1.0375b=4269$ OR $0.08a+0.0375b=269$
62.	a. \$881.25 b. 2.5% c. \$6,250
63.	Equation 1: $a+b+c=15000$ Equation 2: $0.02a+0.05b+0.08c=720$ Equation 3: $b=a+4000$
64.	Equation 1: $b+g+y=19200$ Equation 2: $2b+3g+3.5y=56400$ Equation 3: $g=3b$
65.	Equation 1: $E+P+F=3000$ Equation 2: $1.25E+1.40P+1.75F=4420$ Equation 3: $F=E$
66.	$x=1$
67.	The distance between "10 + x" and "10x" can be expressed as "10x − (10+x)". When simplified, the distance is "9x − 10". For example, if $x=7$, distance is 9(7)−10 or 53. The distance between 17 and 70 is 53.
68.	The perimeter is 15+4+6+10+4+17=56 ft. Use the Pythagorean Theorem to find missing lengths in the figure. 4 ft, 17 ft, 8 ft, 10 ft, 15 ft, 4 ft, 6 ft
69.	The total area is 336 ft^2. Each side of the fence is a rectangle with a height of 6 ft. You can multiply each base by 6ft to find the area of the section.
70.	$\left(\frac{15\cdot 8}{2}\right)+(4\cdot 8)+\left(\frac{6\cdot 8}{2}\right)\rightarrow 60+32+24$ $\rightarrow 116$ ft^2

© Alex Joujan, 2020

71.

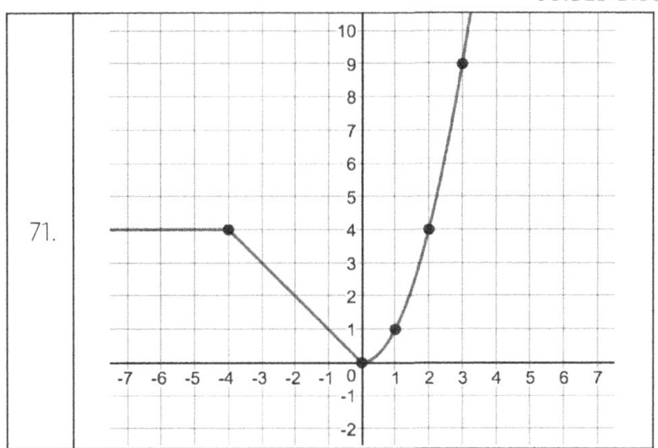

 © Alex Joujan, 2020

 © Alex Joujan, 2020

HOMEWORK & EXTRA PRACTICE SCENARIOS

As you complete scenarios in this part of the book, you will practice what you learned in the guided discovery sections. You will develop a greater proficiency with the vocabulary, symbols and concepts presented in this book. Practice will improve your ability to retain these ideas and skills over longer periods of time.

There is an Answer Key at the end of this part of the book. Check the Answer Key after every scenario to ensure that you are accurately practicing what you have learned. If you struggle to complete any scenarios, try to find someone who can guide you through them.

© Alex Joujan, 2020

CONTENTS

 © Alex Joujan, 2020

Section 1

INTRODUCTION TO RATES SCENARIOS

 © Alex Joujan, 2020

The concept of a rate should not be new to you, but there are many types of scenarios that involve rates. This book will guide you through several topics that involve rates.

1. In the 2012 Nathan's hot dog eating contest, Joey Chestnut won the contest by eating an average of 9 hot dogs every 1.32 minutes.

   a. What was his eating rate, in hot dogs per minute?

   b. How many hot dogs did he eat during the 2012 contest, which lasted 10 minutes?

2. What is the rate shown in the graph? Express the rate as a numerical value and include the units.

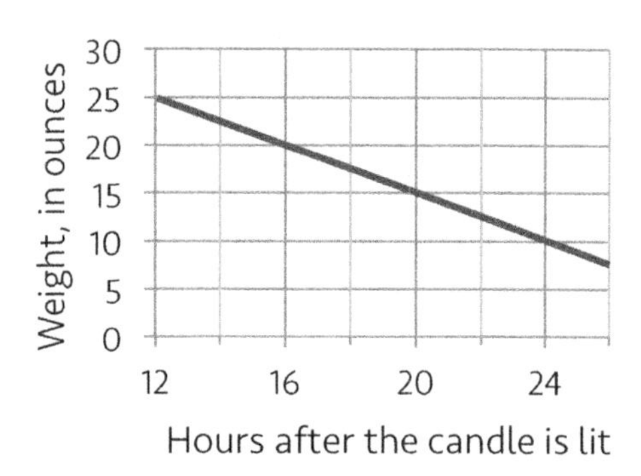

3. A man named Stefaan Engels ran a total of 9,569 miles in one year. A woman named Yolanda Holder walked 521 miles in 15 days. Who averaged a higher daily mileage rate, and by how many miles?

 © Alex Joujan, 2020

4. When egg-laying chickens are raised in henhouses, they are sometimes grouped into cages that contain six hens per cage. In this setting, a traditional cage provides 80 square inches of floor space per bird. If you tiled the floor of one of these cages with sheets of printer paper (measuring 8.5 inches by 11 inches), how many sheets of paper would be needed to cover the entire floor of one cage?

5. ★A cage-free henhouse can hold 500 chickens in 72,000 square inches of total floor space. What is the percent increase in the amount of floor space given to a cage-free chicken, when compared with the space in a traditional cage?

 © Alex Joujan, 2020

Section 2

RATES AND THEIR RECIPROCALS

 © Alex Joujan, 2020

The next group of scenarios will lead you to compare a rate and the reciprocal of that rate in order to see that both of these forms of a rate can be useful.

6. A Lego set costs $50 and it contains 625 pieces.

   a. What is the cost per Lego?

   b. What is the reciprocal of the previous rate?

7. Marley likes to knit scarves and give them to people who live at a local homeless shelter during the cold winter months. She can knit 3 scarves every 6 hours.

   a. What is her knitting rate, in scarves per hour?

   b. What is her knitting rate, in hours per scarf?

8. James types at speed that is close to that of a typical person. He can type 1200 words in 30 minutes.

   a. How many words does James type in one minute?

   b. It takes ______ seconds for James to type one word.

9. Beth paints the walls of a bedroom at a rate of 8 square feet per minute. How long will it take her to paint an entire room if it has four equally sized walls and each wall is 15 feet wide and 8 feet high?

 © Alex Joujan, 2020

Section 3

SCENARIOS INVOLVING WORK RATES AND COMBINED WORK

 © Alex Joujan, 2020

10. At a county fair, you sit in the arts and crafts section and braid bracelets. On average, it takes you 6 minutes to braid an entire bracelet.

    a. What is your braiding rate, measured in bracelets per minute?

    b. How many bracelets would you braid in 20 minutes?

11. A basketball player works through a drill in which she shoots 20 free throws in one minute.

    a. How long does it take her to shoot one free throw, measured in minutes?

    b. How long does it take her to shoot N free throws?

    c. What is the rate at which she shoots free throws, measured in free throws per minute?

12. Guess the missing information in the scenario below. Do not set up and solve an equation.

    One bird can build a nest in 10 hours. Another bird can build a nest in _______ hours. Working together, they can build a nest in 6 hours.

13. Solve each equation below.

    a. $\frac{1}{5}y+\frac{1}{2}y=1$

    b. $\frac{1}{40}x+\frac{1}{60}x=1$

 © Alex Joujan, 2020

14. Grass grows throughout the season, so the lines on athletic fields need to be repainted weekly. An experienced worker can repaint an entire field in 8 hours, while a novice worker will need 12 hours to repaint the field. If they work together, how long will it take them to repaint the field? You might think they would do the entire job in 10 hours since that is halfway between 8 and 12. A better guess would be half of the average of their times, or 5 hours, but this still doesn't turn out to be the actual time. Consider the scenario in smaller portions.

a. What fractional portion of the field does the experienced worker repaint in 1 hour? In 2 hours? In H hours?

b. What portion of the field does the novice worker repaint in H hours?

c. Working together, how much of the field do they repaint in 1 hour?

d. Working together, how long will it take them to repaint the entire field?

15. Your grandparents live in a single-level home, which has no stairs, and you buy them 2 robotic vacuums as gifts. One vacuum is an iRobot Roomba, and it cleans the floor in their entire home in 120 minutes. The other vacuum is a Neato Botvac, which cleans their floors in 90 minutes.

a. What fractional amount of the house can each vacuum clean in 1 minute?

b. Working together, what fractional amount of the floors can the vacuums clean in 2 minutes?

c. How much can the vacuums clean in M minutes?

 © Alex Joujan, 2020

16. If the vacuums in the previous scenario are turned on at the same time, how long will it take them to work together to vacuum the entire home?

17. A farm is divided into seven equally sized fields. A farmer can plow one of these fields in 3 hours. Her son can plow each field in 4 hours.

    a. If they work together, driving separate tractors, how much can they plow in 1 hour?

    b. How much can they plow in 2 hours?

    c. How much can they plow in H hours?

    d. How much can they plow if the farmer works X hours and her son works Y hours?

18. If they work together the entire time, write an equation that you can use to calculate how long it will take the farmer and her son to plow the seven fields, if time is measured in hours.

19. ★Suppose the farmer in the previous scenario wakes up early and starts plowing alone. Her son joins her after 90 minutes, and they work together to finish plowing what remains of the seven fields. How long do they work together?

 © Alex Joujan, 2020

20. Ayush and Sahil are brothers who work at the same drone company. Usually they like to work separately to build drones, but sometimes they work together. When Ayush works alone, he can assemble an entire drone in 100 minutes. When they work together at the same time, their work is modeled by the equation below.

$$\frac{1}{75}m+\frac{1}{100}m=1$$

In this equation, m is the amount of time that it takes them to build 1 drone when they work together.

a. Which brother works at a faster pace?

b. If Ayush and Sahil work together, how many drones can they build in 5 hours?

21. In the winter, Sara and Eli work for a store that sells firewood. Their job is to organize truckloads of chopped firewood into stacks. Eli can stack a truckload in 2 hours. Sara takes twice as long to stack one truckload. If they work together, how many truckloads can they stack in an 8-hour workday?

 © Alex Joujan, 2020

22. Kyra and Cane both work for their dad at his bank on Saturday morning. Their job is to fill containers with bundles of dollar bills. When Kyra and Cane work together, they fill one container in 8 minutes. When Kyra does all of the work by herself, it takes her 14 minutes to fill one container. How long does it take Cane to fill one container if he works by himself?

23. To earn extra money during the summer, you start a lawn mowing business. On average, it takes you 30 minutes to mow a lawn. Fortunately for you, your clients all have lawns that are roughly the same size. You get so busy that you hire one of your friends to work with you.

a. If you work together on each lawn and it takes 18 minutes to mow each lawn, how long would it take your friend to mow one lawn, working alone?

★b. Suppose you add enough clients to keep you both busy for four and a half hours each day, five days a week (this does not include the time spent driving between houses). If you charge $25 per lawn, how much money does your company earn each week? As a matter of review, this is called your revenue.

★c. If you spend an average of $75 per day on gas and equipment repairs costs, how much profit do you earn each week?

★d. If you decrease your weekly costs by 20%, by what percent will your profits increase?

 © Alex Joujan, 2020

Section 4
SCENARIOS INVOLVING MOTION

 © Alex Joujan, 2020

24. Jonas ran 9 miles in three-fourths of an hour. Henry ran 10 miles in 50 minutes. Which runner averaged a faster rate?

25. Eva goes for a long run in the morning to train for a marathon. She maintains an average running speed of 8 miles per hour. She begins her run at 7:30am and does not stop running until she has covered a distance of 34 miles. What time does she stop running?

26. A robber leaves a bank and starts running down the street. A police officer leaves the bank 5 seconds later and starts running after the robber at a speed of 6 meters per second. If it takes the officer 25 seconds to catch the robber, how fast was the robber running?

27. Kelly and Ruth challenged each other to a toy car race. They planned to place their cars at the top of a ramp and let them go at the same time. The cars would then roll down the ramp and keep going until they stopped. At the start of the race, Ruth's car got a head start because Kelly's wheels were stuck and would not roll. Kelly quickly fixed the wheels and then sent her car down the track 4 seconds later. Ruth's car moved at an average speed of 30 cm per second, but Kelly's car had an average speed of 50 cm/sec. If the cars reached the finish line at the same time, how long was the race, in meters?

 © Alex Joujan, 2020

28. James takes his old dog to a field to run around. The dog sees a squirrel and starts running away. James runs 2 meters per second faster than his dog, but the dog is already 10 meters ahead before James starts running. How long does it take for James to catch his dog?

29. Jed and Van are Ruth and Kelly's younger brothers. They think it will be more fun to make their cars collide so they set up a system that makes the cars launch from opposite ends of a room. Jed's launcher makes his car roll at an average speed of 60 cm per second, but Van's launcher is stronger and his car averages 80 cm per second after it leaves the launcher. Their launchers are separated by a distance of 610 centimeters (about 20 feet). When it is time to let the cars go, Van gets distracted and lets his car go 2 seconds after Jed. The cars collide somewhere in the middle of the room. Which car travels a greater distance before they collide and by how many centimeters?

30. Felix and Ron are two brothers who live on opposite coasts. They meet in Kansas to spend a week together. At the end of the week, Felix drives back to his house, but Ron flies home. Ron's flight leaves 3 hours after Felix starts driving. Felix drives at an average speed of 65 miles per hour. If they are travelling in opposite directions and they are 1200 miles apart after Felix has been driving for 5 hours, what is the average flying speed of Ron's plane?

31. After practice one afternoon, the entire team heads to the track for a 60-yard sprinting drill. In one cycle of the drill, you run 60 yards down the field until you reach a line of cones and then you turn around and walk back to the starting point again. One cycle of the drill is completed when you reach the cones and walk back again. If you complete one cycle of the drill in 56 seconds and your walking speed is 1.5 yards per second, what is your running speed during the drill, in yards per second?

 © Alex Joujan, 2020

Section 5
CALCULATING AVERAGE SPEED

 © Alex Joujan, 2020

32. The graph displays data from another one of your bike rides. The entire ride is shown in the graph.

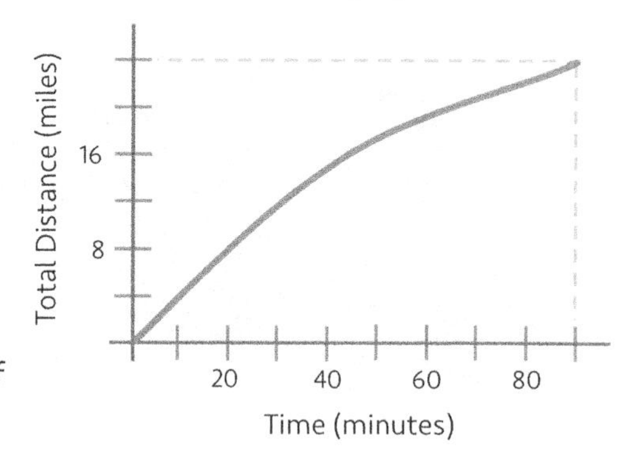

a. Looking at the graph, what was your average speed on this bike ride, in miles per hour?

b. What was your biking speed during the first 20 minutes of your ride?

c. Write a short story that describes how fast you rode during your entire bike ride.

33. During a week in the summer, Jo tracks how many steps she has taken, but she forgot to write down her information a few times. On average, how many steps did she take each day during the week shown below? Use estimation to answer this question.

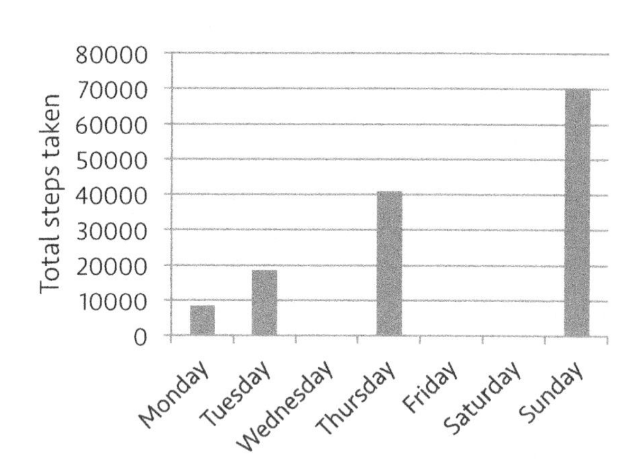

34. Nia drove 300 miles in 4 hours, 180 miles in 3 hours, and 50 miles in 1 hour. What was her average speed for the entire drive?

 © Alex Joujan, 2020

35. If you travel 40 miles in 40 minutes and then slow down and travel another 40 miles in 60 minutes, what is your average speed for the entire trip?

36. When you drive 10 miles to the grocery store, the car maintains an average speed of 30 miles per hour. On the way back home, which is also a 10-mile drive, the car averages a speed of 40 miles per hour. What was the average speed of the entire drive?

37. ★Each lap in a car race is 2.5 miles. A driver averaged a speed of 195 miles per hour during the first 3 laps of the race. If he completed the first two laps in 47.2 seconds, and 45.8 seconds, respectively, how long did it take to complete the third lap?

38. A runner enters a 4-mile race. The runner covers 3 miles in 20 minutes and then speeds up to run the rest of the race in only 6 minutes. What was the runner's average speed for the entire race, in miles per hour?

 © Alex Joujan, 2020

39. The graph to the right displays data from the entire route traveled by a bicyclist during a race.

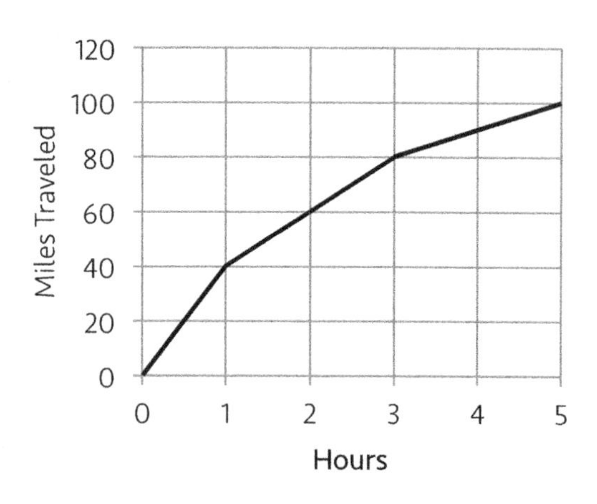

a. What was the average speed of the rider over the course of the entire race?

b. Estimate the rider's maximum speed during the race and support your estimate with numerical values.

40. Vincent has run a 10-mile race in exactly 1 hour, and he has another race coming on Saturday. When the race finally arrives, he feels like he will be able to beat his best time. For the first 6 miles, Vincent is able to maintain an average speed of 12 miles per hour. He gets really tired, though, and he only averages 8 miles per hour for the rest of the race. Did he set a personal record when he finished the race?

 © Alex Joujan, 2020

Section 6

SCENARIOS INVOLVING INTEREST RATES

© Alex Joujan, 2020

41. You invest \$400 in your favorite stock. One year later it is worth \$440. What was the percent increase in the value of the stock?

42. ★You invest \$500 and in 2 years, its value increases to \$800. What was the average yearly percent increase in the value of your investment? This can be difficult to calculate, so be prepared for possible confusion.

43. You open a savings account that earns 2.5% in interest every year. If you deposit \$3,000 in the account when you open it, how much will the account be worth after 5 years?

44. Estelle gives advice to people who want to invest their money. One of her clients wants to invest a total \$20,000 dollars. Estelle splits the money between two accounts. One account is a mutual fund that increases in value by 8% after one year. The other account is a government savings bond that increases in value by 4% after one year. If Estelle's investment decision led her client to increase the value of the initial investment by \$1,360 after one year, how much did Estelle invest in each account?

45. Viv bought two different types of stocks, A and B. Viv invested a total of \$6,000. After 1 year, stock A increased by 4% and stock B increased by 11%. Viv was fascinated to find out that even though the interest rates were different, both of her investments earned the same amount of money in interest that year.

    a. How much money did she invest in each of the two stocks?

    b. How much did Viv earn in interest that year?

 © Alex Joujan, 2020

46. In the previous scenario, how much money would Viv have earned in interest if she could go back in time and switch the amounts that she invested in the 2 stocks?

47. Leigh buys shares of stocks in company F and in company G. The equations below show the relationship between the amount of money she invested and the amount of interest she earned after one year.

Equation 1: $f = g + 2{,}500$
Equation 2: $0.06f + 0.037g = 402.20$

a. How much did Leigh earn in interest after one year?

b. By what percent did the value of company G's stock increase after 1 year?

c. How much money did Leigh invest in company F?

48. Your friend sees that you are interested in learning more about investing so she opens an account for you and helps you track the investment for one year. She tells you that it is wise to diversify your investments so she puts 50% of the money into one account, 30% into a second account, and the rest into a third account. Over the next 12 months, the first account increases by 11%, the second increases by 5%, and the third account decreases by 4%. At the end of one year, the overall investment increased in value by $62. How much money did she invest for you?

 © Alex Joujan, 2020

Section 7
BOOK 2 REVIEW

 © Alex Joujan, 2020

49. A tank is filled with water, but it has a small leak. Water drips out of the tank at a constant rate. A bucket is placed under the leak to see how much water the tank is losing. When the bucket is checked at 11:00am, there are 60 milliliters of water in the bucket. At 11:20am, there are 164 milliliters of water in the bucket.

a. How much water is leaking out of the tank every minute?

b. How many liters of water does the tank lose in a day?

50. Find the slope of the line, given two points on that line.

a. $(-1, 4)$ and $(8, -2)$

b. $(2, 6)$ and $(2, -3)$

51. Find the slope of the line, given two points on that line.

a. $(11, -7)$ and $(-1, -7)$

b. (x_1, y_1) and (x_2, y_2)

 © Alex Joujan, 2020

52. What is the slope of a line that is parallel to $-5y = 15x - 10$?

53. What is the slope of a line that runs perpendicular to $y = -2x + \frac{2}{3}$?

54. What is the Slope-Intercept Form for the equation of a line?

55. Using Slope-Intercept Form, write the equation of the line, given the following information.

a. one point on the line: (6, 2)

slope: $-\frac{1}{3}$

b. one point on the line: (−3, 4)

y-intercept: (0, −11)

 © Alex Joujan, 2020

56. Convert each equation to Slope-Intercept Form.

a. $5x+7y=-21$

b. $y-4=-\frac{1}{2}(x+12)$

57. What is the Standard Form for the equation of a line?

58. Circle the equations that are in Standard Form, where A, B, and C are integers.

a. $-11x+y=17$

b. $0.5x+2.1y=10.4$

c. $-3x+8y=20$

d. $4x-\frac{1}{3}y=2$

e. $y=2x-5$

f. $y-4=-\frac{1}{2}(x+12)$

59. What is the Point-Slope Form for the equation of a line?

 © Alex Joujan, 2020

60. Write an equation, in Point-Slope Form, for the line described by the given information.

a. one point on the line: $(-2, 6)$

slope: $\frac{5}{2}$

b. two points on the line: $(-8, 5)$ and $(4, -1)$

61. Write an equation, in Point-Slope Form, for the line shown in the graph.

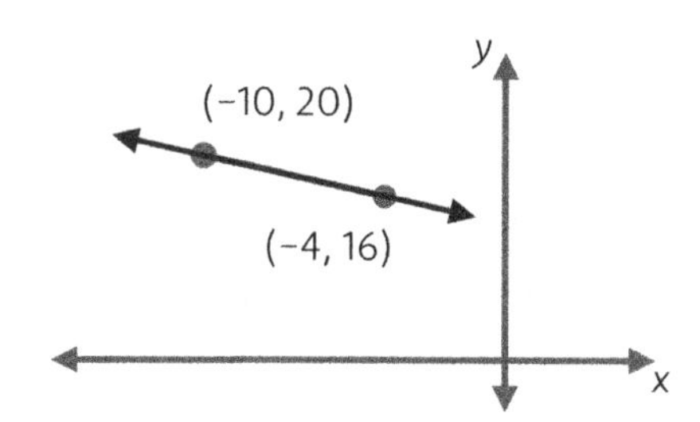

62. Identify the coordinates of the x- and y-intercepts of the line in the previous scenario.

 © Alex Joujan, 2020

63. If a line has an x-intercept of 5, the ordered pair representing the x-intercept is ________.

64. If a line has a y-intercept of −3, the ordered pair representing the y-intercept is ________.

65. When a person exercises, their heartbeat increases. The heartbeats for two people are shown below during a time period when they were exercising. These people have the same heartbeat rates when they are resting. Which person exercised more intensely during the time period shown? How do you know this?

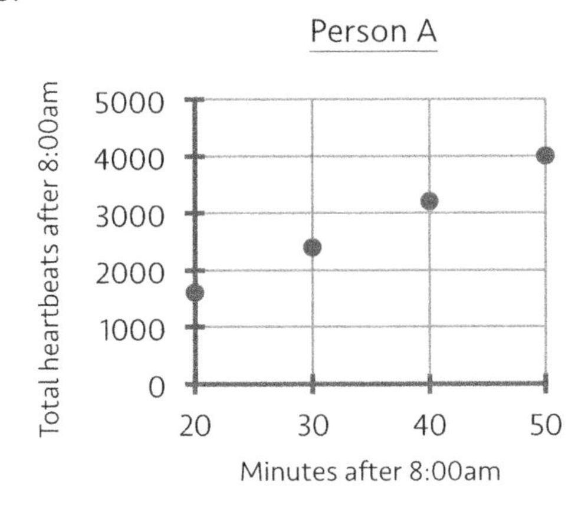

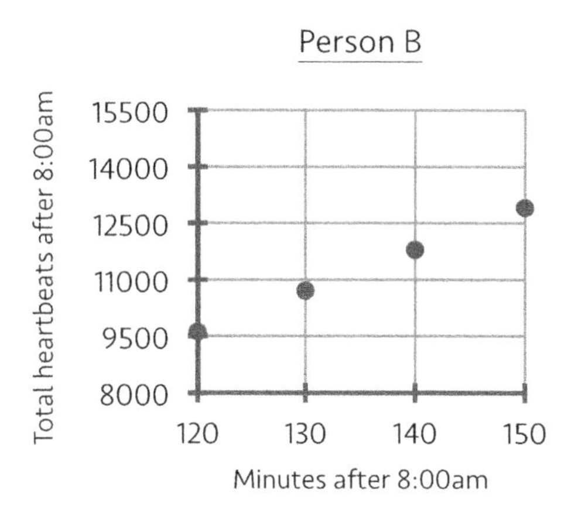

66. The chart to the right shows a person's height at various ages throughout their childhood.

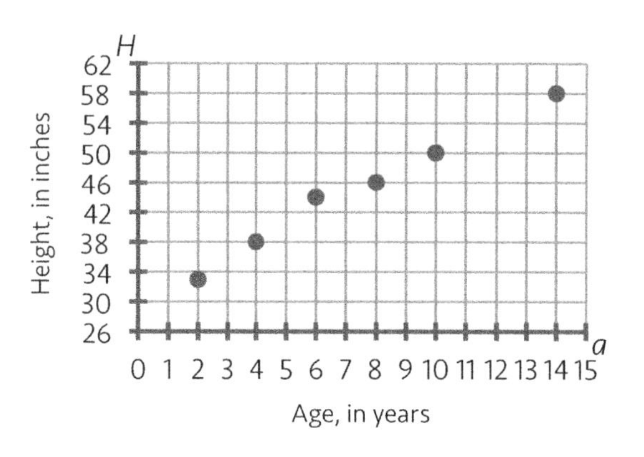

a. Draw an approximate trend line through the scatter plot shown.

b. Find the equation of your trend line to show the relationship between H, the height, and a, the age.

c. What was the average rate of change of the height over the time period shown?

 © Alex Joujan, 2020

67. In the previous scenario, estimate the height of the person when they were born.

68. The following inequalities are written in Standard Form. Rearrange them to write them in Slope-Intercept Form.

a. $14x - 7y < 7$

b. $11x - 22y \geq 66$

69. Did you forget to switch the direction of the inequality in the previous scenario? When you rearrange an inequality, when do you need to reverse the direction of the inequality symbol?

70. Graph each inequality.

a. $y < \frac{1}{3}x + 5$

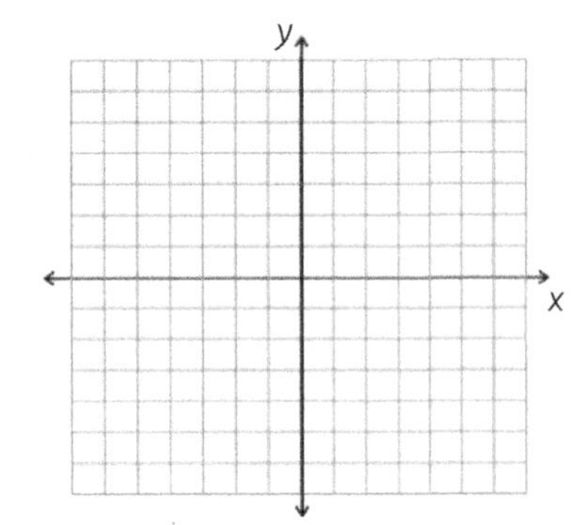

b. $-2y < 2x + 6$

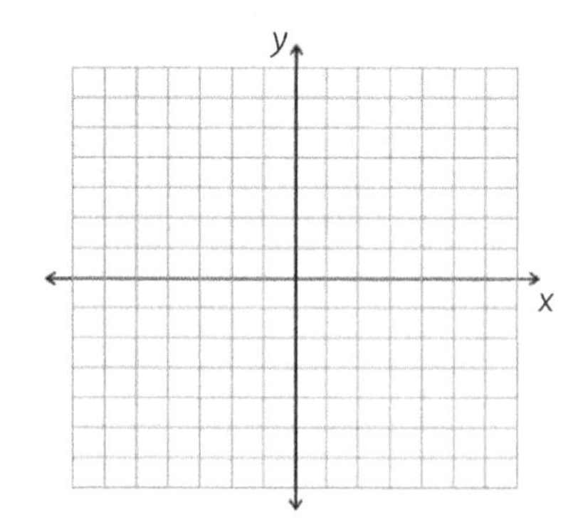

 © Alex Joujan, 2020

71. Graph each inequality.

a. $-3y+6x<15$

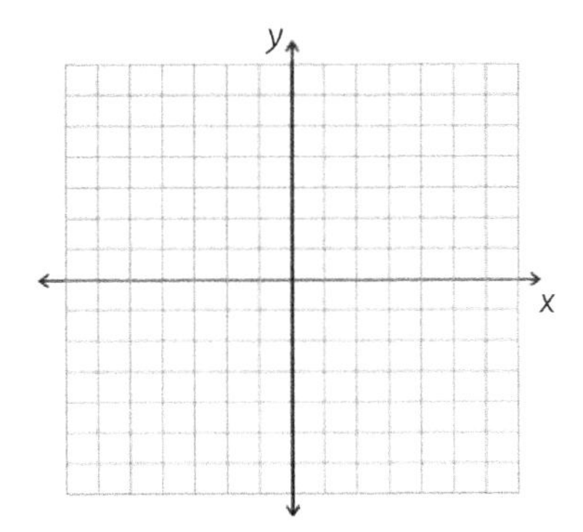

b. $x\geq -4$

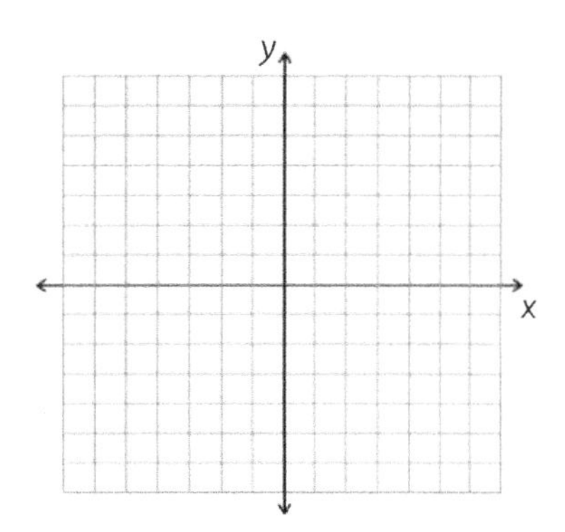

72. When you graph an inequality and you use a dashed line or a solid line for the boundary, what is the reason for choosing either a dashed or a solid line?

 © Alex Joujan, 2020

Section 8
BOOK 3 REVIEW

 © Alex Joujan, 2020

73. When you multiply a number by itself, the result is 25. What was the number?

74. What number equals 16 after it is squared?

When you learn about quadratic functions, you find out that there can be two numbers that make an equation true. This shows up in the previous scenario and in other scenarios you have seen as well.

75. Two quadratic equations are shown below. Solve each equation.

a. $x^2 = 100$

b. $x^2 = -4$

76. Solve each equation.

a. $(x-2)^2 = 9$

b. $-2(x+1)^2 = 10$

77. Fill in the blanks below to make each pair of expression equivalent.

a. $(x+1)^2 = x^2 + 2x + \_\_\_\_$

b. $(x-4)^2 = x^2 + \_\_\_\_ x + \_\_\_\_$

 © Alex Joujan, 2020

78. Write in a third term to make each expression a perfect square trinomial.

a. $x^2 - 10x$ _____

b. $x^2 - 3x$ _____

79. Factor each trinomial above to confirm that it is a perfect square trinomial.

80. Write the Quadratic Formula.

81. Solve each quadratic equation by factoring.

a. $x^2 - 10x + 25 = 0$

b. $3x^2 + 7x = 6$

82. Solve each quadratic equation by Completing the Square.

a. $x^2 - 10x + 2 = 0$

b. $3x^2 - x = 6$

 © Alex Joujan, 2020

83. Solve each quadratic equation by using the Quadratic Formula.

a. $x^2 - 10x + 2 = 0$

b. $3x^2 - x = 6$

84. What is the shape of the graph of $y = mx + b$?

85. What is the shape of the graph of $y = ax^2 + bx + c$?

86. How can you use a parabola's equation to determine if a parabola will open upward or downward?

87. Graph the function below. You must include the vertex, the x-intercepts, and the y-intercept in your plotted points. Plot at least 7 points.

$y = -x^2 + 6x$

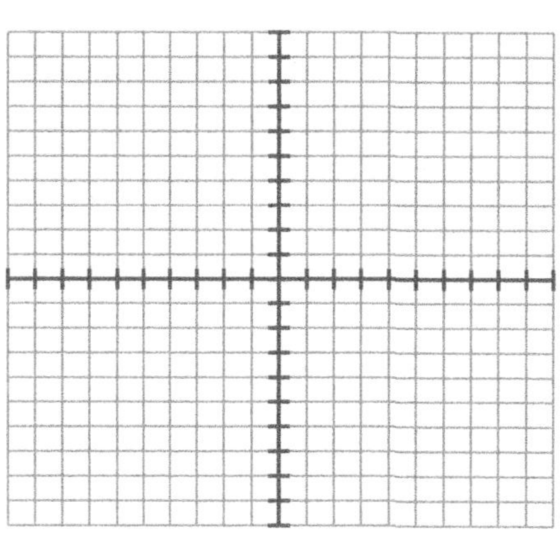

 © Alex Joujan, 2020

88. Graph the function below. You must include the vertex, the x-intercepts, and the y-intercept in your plotted points. Plot at least 7 points.

$$f(x) = -\frac{1}{2}x^2 + 6$$

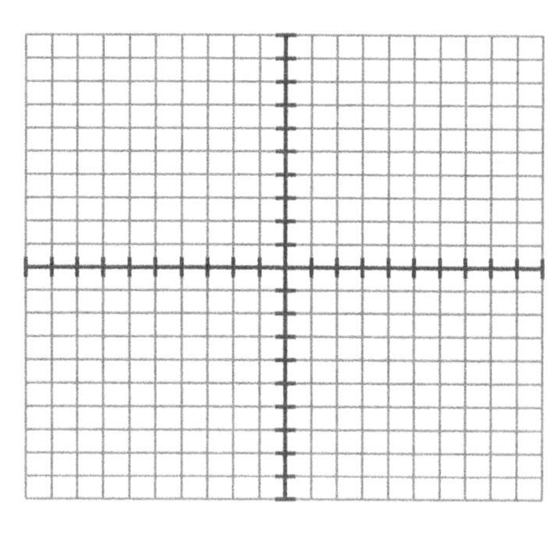

89. A basketball player attempts a shot. The equation for the height, h, of the ball (in feet) as a function of time, t, (in seconds) is given by $h(t) = -16t^2 + 32t + 6$.

a. What is the maximum height of the ball during its trajectory?

b. The ball misses the basket and hits the floor below the net. For how many seconds was the ball in the air before it landed on the floor?

 © Alex Joujan, 2020

90. Fill in the blank. When you find the maximum height of a ball as it flies through the air, you're really just finding the location of the ____________ of a parabola.

91. Fill in the blank. When you calculate how long a ball is in the air before it hits the ground, you're really just looking for the ______________ of a parabola.

92. Determine the x-intercepts of the function $y=x^2-8x-5$.

 © Alex Joujan, 2020

Section 9

BOOK 4 REVIEW

 © Alex Joujan, 2020

93. Simplify each of the following fractions.

a. $\frac{3x(x+4)}{5x(x+4)}$ b. $\frac{(x+1)(x+3)}{(x+1)(x+7)}$ c. $\frac{6x+6}{7x+7}$

94. In the previous scenario, the last fraction can only be simplified after you factor the expressions in the numerator and denominator. After you factor, it becomes clear that there are identical factors in the numerator and denominator. Use this strategy to simplify the following fractions.

a. $\frac{x^2-2x}{x^2-7x+10}$ b. $\frac{6x^2-7x-5}{3x^2+4x-15}$

95. When you simplify a fraction, you remove a disguised form of 1. Write the disguised form of 1 that you removed in each fraction in the previous scenario.

96. What is the simplified form of each fraction shown below?

a. $\frac{x+3}{3+x}$ b. $\frac{3y-8}{8-3y}$ c. $\frac{4-y}{5(y-4)}$ d. $\frac{(y-9)(y-1)}{(1-y)}$

© Alex Joujan, 2020

97. Write the disguised form(s) of 1 in each expression below. Do not multiply the fractions.

a. $\frac{x}{8} \cdot \frac{12}{6x}$

b. $\frac{x(x-3)}{21} \cdot \frac{14}{(x-3)}$

c. $\frac{(x-7)^2}{5x+20} \cdot \frac{x+4}{(x-7)(x-7)}$

98. Simplify as much as you can.

a. $\frac{x^2-6x}{x+4} \cdot \frac{2x^2+8x}{x-6}$

b. $\frac{x^2-49}{x^2+6x-7} \cdot \frac{5x^2-4x-1}{7-x}$

When you divide expressions, it is useful to rewrite the division as a multiplication scenario. In other words, think of "divide" as "multiply by the reciprocal" and then carry on with the familiar process of multiplying fractions and simplifying the result.

99. Simplify as much as you can.

a. $\frac{-7x+35}{x^3+5x^2} \div \frac{25-x^2}{x^3+10x^2+25x}$

b. $\frac{4y^2-1}{4y^2+12y} \div \frac{2y^2-15y-8}{y^2-5y-24}$

© Alex Joujan, 2020

When you compare multiplying and dividing fractions, division is related to multiplication. When a scenario requires dividing by a fraction, you can change the division to multiplying by the reciprocal. Similarly, when you compare adding and subtracting fractions, subtraction is related to addition.

100. Simplify as much as you can.

a. $\frac{5+x}{2x}+\frac{3-4x}{x}$

b. $\frac{5-x}{x-2}-\frac{2x-1}{x-2}$

101. Add the fractions. Simplify as much as you can.

a. $\frac{p+2}{p+5}+\frac{p-5}{p^2+10p+25}$

b. $\frac{8}{t-3}+\frac{5}{t}$

102. Subtract the fractions. Simplify as much as you can.

a. $\frac{5}{y-1}-\frac{3}{y+2}$

b. $\frac{4}{x^2-25}-\frac{x+2}{x^2+x-20}$

 © Alex Joujan, 2020

103. Solve the equation.

$$\frac{x}{4}+\frac{x+7}{3}=7$$

104. You can make the previous equation easier to solve by eliminating the fractions. To do this, multiply both sides of the equation by 12 (the least common multiple of 3 and 4). You can practice clearing the fractions in each equation shown below. Do not solve each equation. Instead, write the expression that will clear the fractions if you multiply both sides of the equation by that expression.

a. $\frac{6}{3x}+\frac{-2}{2}=\frac{11}{3}$ b. $x-\frac{4}{x}=\frac{1}{2}$ c. $\frac{3}{x+1}+\frac{x-4}{2}=\frac{1}{x+1}$

105. Clear the fractions in each equation in the previous scenario. Do not solve the equation.

106. Solve each equation.

a. $\frac{10}{y}+\frac{2}{5}=\frac{6}{y}$ b. $\frac{-2}{x+3}-\frac{6}{x}=1$

 © Alex Joujan, 2020

107. Check your solution for each of the previous equations to confirm that your solution makes the original equation true.

108. Solve the equation.

$$\frac{2}{x^2+4x-5}-\frac{10}{x+5}=\frac{6}{x-1}$$

 © Alex Joujan, 2020

Section 10

BOOK 5 REVIEW

109. Two brothers apply for a job at a bakery. The job involves baking cookies, so the bakery owner asks them to come in and bake cookies on separate days. Their results are shown in the table. If they both baked good-tasting cookies, which brother was hired for the job?

Name	Time spent at the bakery	Cookies baked
Adnan	45 minutes	216 cookies
Omer	1.2 hours	360 cookies

110. Suppose Omer and Adnan are both hired to work at the bakery. The bakery sets a goal to make 3,000 cookies per day. How many hours do they need to spend at the bakery every day to reach the bakery's goal? The brothers work together and maintain the same baking rates from the previous scenario.

111. Solve the equation.

$$\frac{1}{8}t+\frac{1}{5}t=1$$

 © Alex Joujan, 2020

112. Kylie's team can build a house in 8 days. Dan's team can build a house in 5 days. If they work together, how long will it take them to build a house?

113. Bill and Gus are co-owners of an apple orchard. Bill can harvest all of the apples in 12 days. Gus can harvest all of the apples in 10 days. How long does it take hem to harvest all of the apples when they work together?

114. Joyce walks 4.08 miles in four-fifths of an hour. Dale walks 3.75 miles in 45 minutes. Which person walks farther in an hour? How much farther does that person walk, measured in feet?

115. You find two ways to travel between two cities. You can take a train or a plane. The train maintains an average speed of 60 miles per hour. The plane flies at an average speed of 310 miles per hour and it will get you to your destination in 2 hours less than the train. What is the distance between the two cities you want to travel between?

 © Alex Joujan, 2020

116. You drive 10 miles in 20 minutes. You then increase your speed and drive 20 more miles in 20 minutes. Finally, you drive 32 more miles in 20 minutes. What was your average speed for the entire trip?

117. If you walk a distance of 1 mile for 10 minutes and then run at a speed of 12 miles per hour for 20 minutes. What was your average speed?

118. You put $500 in account that earns 4% interest every year. You also put some money in a second account that each 3% interest every year. After one year, you earned a total of $41 in interest.

   a. Which account has a higher value of one year?

   b. How much did you invest in the account that earns 3% interest?

 © Alex Joujan, 2020

119. You have saved \$4,000, but you don't need to spend it yet so you look for ways to earn interest on this money. Since you have heard that it is a good idea to diversify your investments, you pick two separate accounts. One account earns 2% interest. Another account does not have a fixed interest rate. Instead, it has a variable interest rate that depends on how the stock market performs that year. You divide your money evenly between the two accounts. At the end of one year, you earned \$100 in interest. By what percent did the value of the variable interest rate account increase after that year?

120. A group of people who are members of their city council are looking for ways to save up for future projects that will improve their city. They agree to invest a total of \$1,000,000 in two types of funds. After one year, the two funds increased in value by 4.5% and 6%, respectively and their initial investment is worth \$1,056,250. How much did they invest in each account?

 © Alex Joujan, 2020

Section 11

CUMULATIVE REVIEW

© Alex Joujan, 2020

121. Which shape has a larger area?

Shape #1: a square with a perimeter of 28cm

Shape #2: a circle with a radius of 4cm

122. A square has an area of 25 square inches. If you double each of the side lengths, by what percent will the area increase?

123. Solve the equation.

$$\frac{-x+2}{x^2-x}=\frac{x}{x-1}-\frac{4}{2x-2}$$

124. Simplify the expression.

$$\frac{6}{x^2-4}\div\frac{3}{x-2}+\frac{3}{x}$$

 © Alex Joujan, 2020

125. Solve each inequality and graph your solution on a number line.

a. $x+1<5$

b. $4-2x \le 16$

126. Graph the inequality on the Cartesian plane provided.

$2x-8y<40$

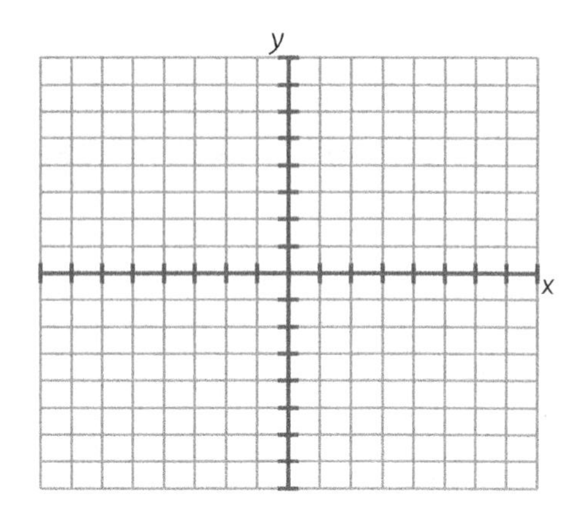

127. A function is defined by the equation $f(x)=2-3x$. A second function, g(x), is shown in the graph. Use the functions to find the value of each expression below.

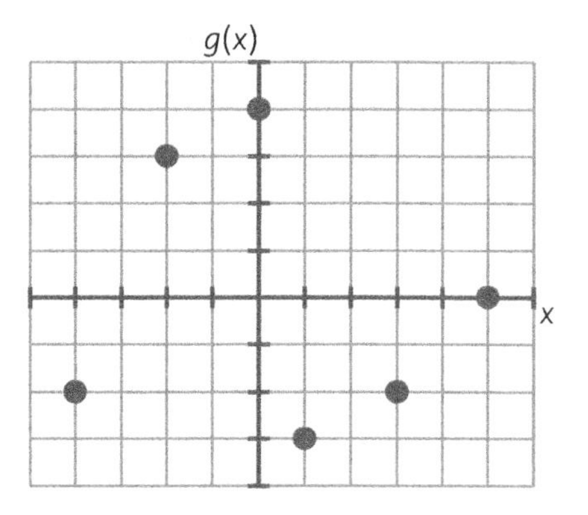

a. $f(-2)$

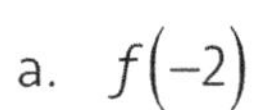

b. $g(0)$

c. the value of x when $g(x)=-2$

d. $f[g(1)]$

128. A function is defined as shown below. Graph the function.

$y=-2x+8$ if $x>2$

$y=x^2$ if $-2 \le x \le 2$

$y=2x+8$ if $x<-2$

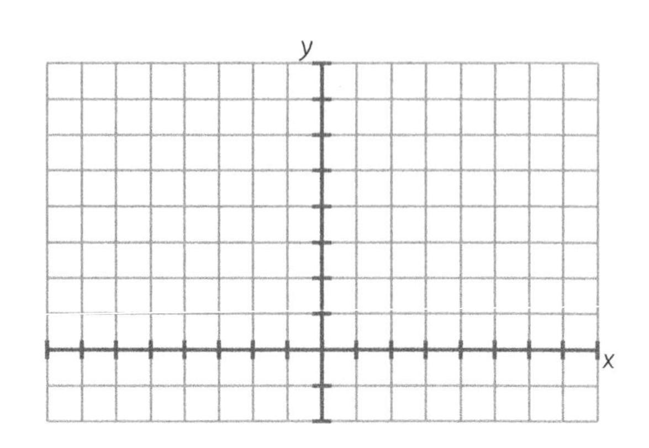

 © Alex Joujan, 2020

Section 12

ANSWER KEY

#	Answer
1.	a. 6.82 hot dogs/min b. 68
2.	a decrease of 1.25 ounces per hour
3.	Yolanda, by about 8.5 miles per day
4.	A little more than 5 sheets (≈5.13) area of floor: 480 sq. in. area of sheet of paper: 93.5 sq. in.
5.	80% more (144 vs. 80)
6.	a. \$0.08/piece b. 12.5 pieces/dollar
7.	a. 0.5 scarf per hr. b. 2 hrs. per scarf
8.	a. 40 words per minute b. 1.5
9.	1 hour
10.	a. $\frac{1}{6}$ of a bracelet per minute b. $\frac{1}{6}(20)=\frac{20}{6}=3\frac{1}{3}$ bracelets
11.	a. $\frac{1}{20}$ b. $\frac{N}{20}$ c. 20 per minute
12.	15 hours
13.	a. $\frac{2}{10}y+\frac{5}{10}y=1\rightarrow\frac{7}{10}y=1\rightarrow y=\frac{10}{7}$ or $1\frac{3}{7}$ b. $\frac{3}{120}x+\frac{2}{120}x=1\rightarrow\frac{5}{120}x=1\rightarrow x=24$
14.	a. $\frac{1}{8}, \frac{1}{4}, \frac{1}{8}H$ b. $\frac{1}{12}H$ c. $\frac{1}{8}+\frac{1}{12}\rightarrow\frac{5}{24}$ d. solve $\frac{1}{8}H+\frac{1}{12}H=1$; $H=\frac{24}{5}$ (4.8 hrs)
15.	a. $\frac{1}{120}$ (iRobot) and $\frac{1}{90}$ (Neato) b. $2\left(\frac{1}{120}+\frac{1}{90}\right)\rightarrow\frac{7}{180}$ c. $M\left(\frac{1}{120}+\frac{1}{90}\right)\rightarrow\frac{7}{360}M$
16.	$51\frac{3}{7}$ min
17.	a. $\frac{1}{3}+\frac{1}{4}\rightarrow\frac{4}{12}+\frac{3}{12}\rightarrow\frac{7}{12}$ of a field b. $\frac{1}{3}(2)+\frac{1}{4}(2)$ or $1\frac{1}{6}$ fields c. $\frac{1}{3}(h)+\frac{1}{4}(h)$ or $\frac{7}{12}h$ fields d. $\frac{1}{3}X+\frac{1}{4}Y$
18.	$\frac{1}{3}h+\frac{1}{4}h=7$
19.	solve: $\frac{1}{3}(h+1.5)+\frac{1}{4}h=7$; $t=11\frac{1}{7}$ hours
20.	a. Sahil can build a drone in 25 fewer minutes than his brother Ayush b. They can build exactly 7 drones in 5hr. $\frac{1}{75}\cdot\frac{4}{4}m+\frac{1}{100}\cdot\frac{3}{3}m=1\rightarrow\frac{4}{300}m+\frac{3}{300}m=1$ $\frac{7}{300}m=1\rightarrow\frac{300}{7}\cdot\frac{7}{300}m=1\cdot\frac{300}{7}$ $\rightarrow m=\frac{300}{7}$ minutes They build 1 drone in $\frac{300}{7}$ min. In 300 mins, they would build 7 drones.
21.	They can stack 6 truckloads in 8 hours. solve: $\frac{1}{2}h+\frac{1}{4}h=1$ $h=1\frac{1}{3}$ hours to stack 1 truckload
22.	$18\frac{2}{3}$ minutes solve: $\frac{1}{14}(8)+\frac{1}{C}(8)=1\rightarrow C=18\frac{2}{3}$
23.	a. 45 min b. \$1,875 c. \$1,500 d. 5%
24.	Both runners average a distance of 1 mile in 5 minutes so they are running at the same speed.
25.	11:45am (she runs for 4.25 hours)
26.	5 meters per second They run the same distance when they leave the bank so you can make RxT for the officer equal to RxT for the robber.

 © Alex Joujan, 2020

	solve: 6x25 = Rx30
27.	The race was 3 meters long (300 cm). solve: $30t = 50(t - 4) \to t = 10$ Ruth's car starts earlier but they travel the same distance so you can write expressions for the separate distances and make them equal to each other.
28.	James will catch his dog in 5 seconds. He gets 2 meters closer to his dog every second and the dog starts 10 meters ahead of him.
29.	Jed's car travels 50 more centimeters. solve: $60t + 80(t - 2) = 610 \to t = 5.5$ sec The cars start 610cm apart so you can write expressions for their 2 distances and make them equal 610. Jed's car travels 330cm (60x5.5) Van's car travels 280cm (80x3.5)
30.	437.5 miles per hour Felix drives 325 miles in 5 hours. Since they are 1200 miles apart, Ron has flown 875 miles in 2 hours.
31.	3.75 yards per second
32.	a. 16 mph b. 24 mph c. Answers may vary
33.	70,000 after 7 days = 10,000 steps/day
34.	Total distance: 530 miles Total time: 8 hours Average speed = 530÷8 = 66.25 mph
35.	usual mistake: (40+60)÷2 = 50mph actual result: 80mi÷100min = 48 mph
36.	usual mistake: (30+40)÷2 = 35mph actual result: 20mi÷35min = $34\frac{2}{7}$ mph
37.	The third lap was about 45.5 seconds. It takes about 138.5 seconds to travel 3 laps (7.5 miles) at a speed of 195 mph.
38.	4 miles÷26 min=Approx. 9.23 mph
39.	a. 100÷5= 20 mph b. between hours 0 and 1, the rate is approx. 40 mph → this part of the graph has the steepest slope
40.	It is impossible to tell. He completed the race in 1 hour again, but he may have run slightly faster than his best time if the average speeds are rounded to the nearest whole number. He also may have run slightly slower than his best time.
41.	10%
42.	About 26.5%
43.	About $3,394
44.	Equation 1: A + B = 20,000 Equation 2: .08A + .04B = 1,360

	$14,000 in the mutual fund $6,000 in the gov. savings bond
45.	a. Equation 1: A + B = 6,000 Equation 2: .04A = 0.11B A: $4,400 B: $1,600 b. $352
46.	.04(1600) + .11(4400) = $548 in interest
47.	a. $402.20 b. 3.7% c. $5,100
48.	Let T = total amount invested 1st account = $0.5T$; 2nd account: $0.3T$; 3rd account = $0.2T$ $.11(.5T)+.05(.3T)-.04(.2T)=62 \to T=\$1{,}000$
49.	a. 5.2 mL per minute. The bucket loses 104 milliliters in 20 minutes. b. $5.2\times60\times24=7488\text{mL}\approx7.5\text{L}$ in a day
50.	a. $\frac{-2-4}{8-(-1)}=\frac{-6}{9}=-\frac{2}{3}$ b. $\frac{-3-6}{2-2}=\frac{-9}{0}\to$ undefined
51.	a. $\frac{-7-(-7)}{-1-11}=\frac{0}{-12}=0$ b. $\frac{y_2-y_1}{x_2-x_1}$
52.	$\frac{-5y}{-5}=\frac{15x-10}{-5}\to y=-3x+2\to$ slope is -3
53.	Perpendicular lines have opposite reciprocal slopes. The opposite reciprocal of -2 is $\frac{1}{2}$.
54.	$y = mx + b$
55.	a. $y=-\frac{1}{3}x+b\to$ let x=6 and y=2 $2=-\frac{1}{3}(6)+b\to2=-2+b\to b=4$ $y=-\frac{1}{3}x+4$ b. $y=mx-11\to$ let x=−3 and y=4 $4=m(-3)-11\to15=-3m\to m=-5$ $y=-5x-11$
56.	a. $7y=-5x-21\to y=-\frac{5}{7}x-3$ b. $y-4=-\frac{1}{2}x-6\to y=-\frac{1}{2}x-2$
57.	$Ax + By = C$
58.	Circle a. and c.
59.	$y-y_1=m(x-x_1)$
60.	a. $y-6=\frac{5}{2}(x+2)$

 © Alex Joujan, 2020

	b. Find the slope: $\frac{-1-5}{4-(-8)}=\frac{-6}{12}=-\frac{1}{2}$ Use either point to write the equation: $y-5=-\frac{1}{2}(x+8)$ or $y+1=-\frac{1}{2}(x-4)$
61.	Find the slope: $\frac{16-20}{-4-(-10)}=\frac{-4}{6}=-\frac{2}{3}$ Use either point to write the equation: $y-16=-\frac{2}{3}(x+4)$ or $y-20=-\frac{2}{3}(x+10)$
62.	$y-16=-\frac{2}{3}(x+4)\rightarrow y-16=-\frac{2}{3}x-\frac{8}{3}$ $y=-\frac{2}{3}x-\frac{8}{3}+\frac{48}{3}\rightarrow y=-\frac{2}{3}x+\frac{40}{3}$ *y*-intercept: $\left(0,\ \frac{40}{3}\right)$ or $\left(0,\ 13\frac{1}{3}\right)$ To find the *x*-intercept, let $y=0$. $0=-\frac{2}{3}x+\frac{40}{3}\rightarrow\frac{2}{3}x=\frac{40}{3}\rightarrow x=\frac{40}{3}\cdot\frac{3}{2}$ *x*-intercept: $(20,\ 0)$
63.	$(5,\ 0)$
64.	$(0,\ -3)$
65.	Person B exercised more intensely. Person A had a total heartbeat increase of around 2500 beats in 30 minutes (≈80 beats/min). Person B had a total heartbeat increase of around 3000 beats in 30 minutes (≈100 beats/min).
66.	a. b. $H=2a+30$ c. about 2 inches/year
67.	About 30 inches (the *H*-intercept)
68.	a. $-7y<-14x+7\rightarrow y>2x-1$ b. $-22y\geq-11x+66\rightarrow y\leq\frac{1}{2}x-3$
69.	The direction of the inequality changes when you multiply or divide both sides of the inequality by a negative number.

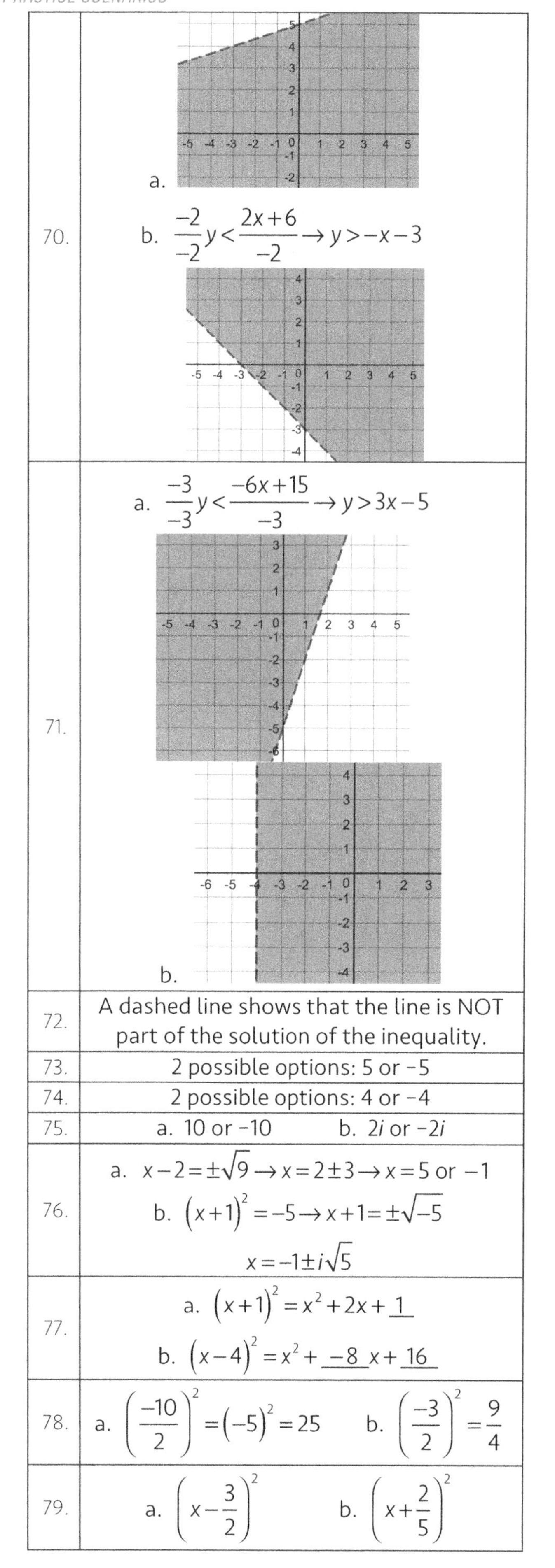

70.	a. b. $\frac{-2}{-2}y<\frac{2x+6}{-2}\rightarrow y>-x-3$
71.	a. $\frac{-3}{-3}y<\frac{-6x+15}{-3}\rightarrow y>3x-5$ b.
72.	A dashed line shows that the line is NOT part of the solution of the inequality.
73.	2 possible options: 5 or −5
74.	2 possible options: 4 or −4
75.	a. 10 or −10 b. $2i$ or $-2i$
76.	a. $x-2=\pm\sqrt{9}\rightarrow x=2\pm3\rightarrow x=5$ or -1 b. $(x+1)^2=-5\rightarrow x+1=\pm\sqrt{-5}$ $x=-1\pm i\sqrt{5}$
77.	a. $(x+1)^2=x^2+2x+\underline{1}$ b. $(x-4)^2=x^2+\underline{-8}\,x+\underline{16}$
78.	a. $\left(\frac{-10}{2}\right)^2=(-5)^2=25$ b. $\left(\frac{-3}{2}\right)^2=\frac{9}{4}$
79.	a. $\left(x-\frac{3}{2}\right)^2$ b. $\left(x+\frac{2}{5}\right)^2$

 © Alex Joujan, 2020

80.	$x=\dfrac{-b\pm\sqrt{b^2-4ac}}{2a}$
81.	a. $(x-5)(x-5)=0\rightarrow x=5$ b. $3x^2+7x-6=0\rightarrow(3x-2)(x+3)$ $x=\dfrac{2}{3}$ or -3
82.	a. $x^2-10x=-2\rightarrow x^2-10x+25=-2+25$ $(x-5)^2=23\rightarrow x-5=\pm\sqrt{23}\rightarrow x=5\pm\sqrt{23}$ b. $\dfrac{3x^2-x}{3}=\dfrac{6}{3}\rightarrow x^2-\dfrac{1}{3}x=2$ $\rightarrow x^2-\dfrac{1}{3}x+\dfrac{1}{36}=2+\dfrac{1}{36}$ $\rightarrow x^2-\dfrac{1}{3}x+\dfrac{1}{36}=\dfrac{73}{36}\rightarrow\left(x-\dfrac{1}{6}\right)^2=\dfrac{73}{36}$ $\rightarrow x-\dfrac{1}{6}=\pm\sqrt{\dfrac{73}{36}}\rightarrow x=\dfrac{1}{6}\pm\dfrac{\sqrt{73}}{6}$
83.	a. $x=\dfrac{-(-10)\pm\sqrt{(-10)^2-4(1)(2)}}{2(1)}$ $x=\dfrac{10\pm\sqrt{92}}{2}\rightarrow x=\dfrac{10\pm2\sqrt{23}}{2}$ $\rightarrow x=5\pm\sqrt{23}$ b. change equation to $3x^2-x-6=0$ $x=\dfrac{1\pm\sqrt{(-1)^2-4(3)(-6)}}{2(3)}\rightarrow x=\dfrac{1\pm\sqrt{73}}{6}$
84.	a line
85.	a parabola
86.	In the equation $y=ax^2+bx+c$, if "a" is positive, the parabola opens upward, like a smile. If a<0, the parabola opens downward, like a frown.
87.	

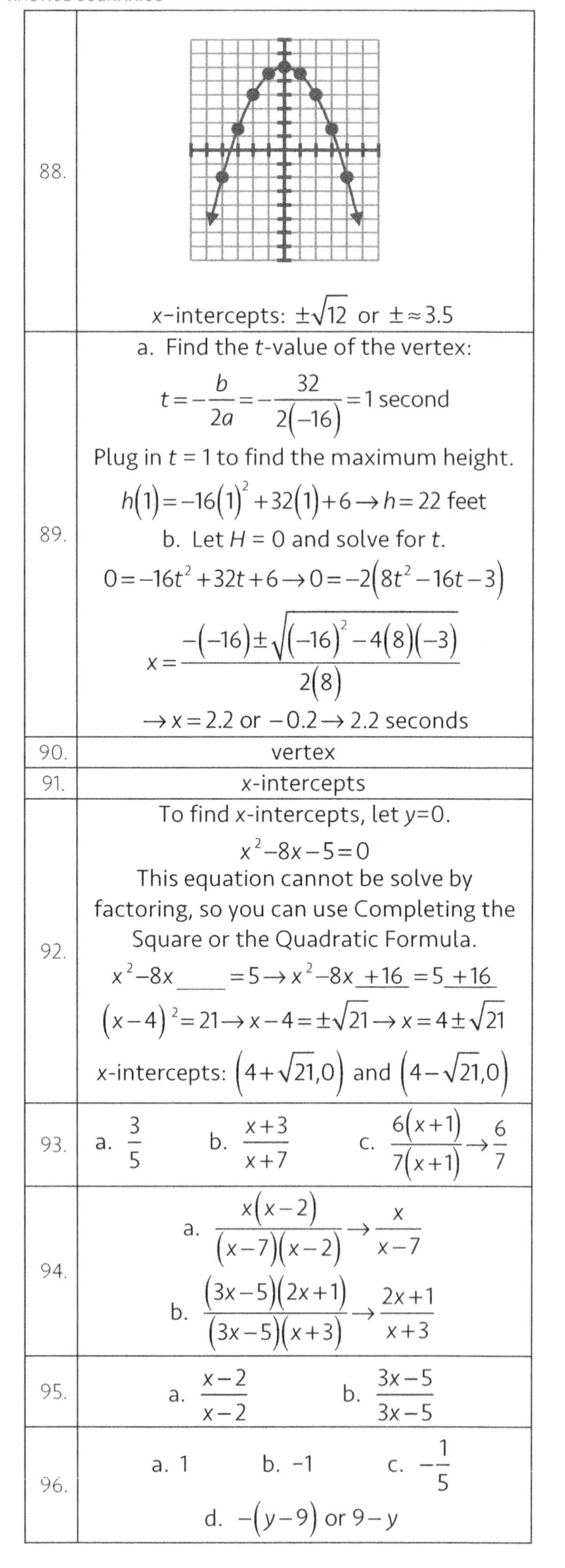

88.	x-intercepts: $\pm\sqrt{12}$ or $\pm\approx3.5$
89.	a. Find the t-value of the vertex: $t=-\dfrac{b}{2a}=-\dfrac{32}{2(-16)}=1$ second Plug in t = 1 to find the maximum height. $h(1)=-16(1)^2+32(1)+6\rightarrow h=22$ feet b. Let H = 0 and solve for t. $0=-16t^2+32t+6\rightarrow 0=-2(8t^2-16t-3)$ $x=\dfrac{-(-16)\pm\sqrt{(-16)^2-4(8)(-3)}}{2(8)}$ $\rightarrow x=2.2$ or $-0.2\rightarrow 2.2$ seconds
90.	vertex
91.	x-intercepts
92.	To find x-intercepts, let y=0. $x^2-8x-5=0$ This equation cannot be solve by factoring, so you can use Completing the Square or the Quadratic Formula. $x^2-8x\underline{\quad}=5\rightarrow x^2-8x\underline{+16}=5\underline{+16}$ $(x-4)^2=21\rightarrow x-4=\pm\sqrt{21}\rightarrow x=4\pm\sqrt{21}$ x-intercepts: $(4+\sqrt{21},0)$ and $(4-\sqrt{21},0)$
93.	a. $\dfrac{3}{5}$ b. $\dfrac{x+3}{x+7}$ c. $\dfrac{6(x+1)}{7(x+1)}\rightarrow\dfrac{6}{7}$
94.	a. $\dfrac{x(x-2)}{(x-7)(x-2)}\rightarrow\dfrac{x}{x-7}$ b. $\dfrac{(3x-5)(2x+1)}{(3x-5)(x+3)}\rightarrow\dfrac{2x+1}{x+3}$
95.	a. $\dfrac{x-2}{x-2}$ b. $\dfrac{3x-5}{3x-5}$
96.	a. 1 b. -1 c. $-\dfrac{1}{5}$ d. $-(y-9)$ or $9-y$

 © Alex Joujan, 2020

97.	a. $\frac{12x}{12x}$ b. $\frac{7(x-3)}{7(x-3)}$ c. $\frac{(x+4)(x-7)^2}{(x+4)(x-7)^2}$
98.	a. $\frac{x(x-6)}{x+4}\cdot\frac{2x(x+4)}{x-6}\to 2x^2$ b. $\frac{(x+7)(x-7)}{(x+7)(x-1)}\cdot\frac{(5x+1)(x-1)}{7-x}\to -(5x+1)$
99.	a. $\frac{-7(x-5)}{x^2(x+5)}\cdot\frac{x(x+5)^2}{(5-x)(5+x)}\to\frac{7}{x}$ b. $\frac{(2y+1)(2y-1)}{4y(y+3)}\cdot\frac{(y-8)(y+3)}{(2y+1)(y-8)}\to\frac{2y-1}{4y}$
100.	a. $\frac{5+x}{2x}+\frac{2(3-4x)}{2(x)}\to\frac{11-7x}{2x}$ b. $\frac{5-x-(2x-1)}{x-2}\to\frac{-3x+6}{x-2}$ $\to\frac{-3(x-2)}{x-2}\to -3$
101.	a. $\frac{(p+2)(p+5)}{(p+5)(p+5)}+\frac{p-5}{(p+5)^2}$ $\to\frac{p^2+7p+10+p-5}{(p+5)^2}\to\frac{p^2+8p+5}{(p+5)^2}$ b. $\frac{8t}{t(t-3)}+\frac{5(t-3)}{t(t-3)}\to\frac{13t-15}{t(t-3)}$
102.	a. $\frac{5(y+2)}{(y-1)(y+2)}-\frac{3(y-1)}{(y+2)(y-1)}$ $\to\frac{5y+10}{D}-\frac{3y-3}{D}\to\frac{2y+13}{(y-1)(y+2)}$ b. $\frac{4}{(x+5)(x-5)}-\frac{x+2}{(x+5)(x-4)}$ $\frac{4(x-4)}{(x+5)(x-5)(x-4)}-\frac{(x+2)(x-5)}{(x+5)(x-4)(x-5)}$ $\frac{4x-16}{D}-\frac{x^2-3x-10}{D}\to\frac{-x^2+7x-6}{D}$ $\to\frac{-(x^2-7x+6)}{D}\to\frac{-(x-6)(x-1)}{(x+5)(x-5)(x-4)}$
103.	$12\cdot\left(\frac{x}{4}+\frac{x+7}{3}\right)=7\cdot 12\to 3x+4x+28=84$ $\to 7x=56\to x=8$
104.	a. $6x$ b. $2x$ c. $2(x+1)$

105.	a. $6x\cdot\left(\frac{6}{3x}+\frac{-2}{2}\right)=\left(\frac{11}{3}\right)\cdot 6x\to 12-6x=22x$ b. $2x\cdot\left(x-\frac{4}{x}\right)=\left(\frac{1}{2}\right)\cdot 2x\to 2x^2-8=x$ c. $2(x+1)\cdot\left(\frac{3}{x+1}+\frac{x-4}{2}\right)=\left(\frac{1}{x+1}\right)\cdot 2(x+1)$ $\to 6+(x-4)(x+1)=2$
106.	a. $5y\cdot\left(\frac{10}{y}+\frac{2}{5}\right)=\left(\frac{6}{y}\right)\cdot 5y$ $50+2y=30\to 2y=-20\to y=-10$ b. $x(x+3)\left(\frac{-2}{x+3}-\frac{6}{x}\right)=1\cdot x(x+3)$ $-2x-6x-18=x^2+3x\to 0=x^2+11x+18$ $0=(x+2)(x+9)\to x=-2,-9$
107.	a. $\frac{10}{(-10)}+\frac{2}{5}=\frac{6}{(-10)}\to -1+\frac{2}{5}=-\frac{3}{5}$ b. $x=-2\to\frac{-2}{-2+3}-\frac{6}{-2}=1\to -2+3=1$ $x=-9\to\frac{-2}{-9+3}-\frac{6}{-9}=1\to\frac{1}{3}+\frac{2}{3}=1$
108.	$\frac{2}{(x+5)(x-1)}-\frac{10}{x+5}\cdot\frac{(x-1)}{(x-1)}=\frac{6}{(x-1)}\cdot\frac{(x+5)}{(x+5)}$ $\frac{2}{D}-\frac{10x-10}{D}=\frac{6x+30}{D}\to D=(x+5)(x-1)$ $D\cdot\left(\frac{2}{D}-\frac{10x-10}{D}\right)=\left(\frac{6x+30}{D}\right)\cdot D$ $2-(10x-10)=6x+30$ $-10x+12=6x+30\to x=-\frac{9}{8}$
109.	Omer was hired because he baked cookies at a faster rate. Adnan: 288 cookies per hour Omer: 300 cookies per hour
110.	Together, the brothers can bake 588 cookies per hour. The brothers can bake 3,000 cookies in about 5.1 hours.
111.	$40\cdot\left(\frac{1}{8}t+\frac{1}{5}t\right)=1\cdot 40\to 5t+8t=40$ $13t=40\to t=\frac{40}{13}$ or $3\frac{1}{13}$
112.	$3\frac{1}{13}$ days To answer this question, you can solve

© Alex Joujan, 2020

	the equation in the previous scenario.
113.	Solve $\frac{1}{12}d+\frac{1}{10}d=1$. $60\cdot\left(\frac{1}{12}d+\frac{1}{10}d\right)=1\cdot60\rightarrow5d+6d=60$ $11d=60\rightarrow t=\frac{60}{11}$ or $5\frac{6}{11}$ days
114.	$4.08\div\frac{4}{5}\rightarrow4.08\div0.8\rightarrow5.1$ miles/hour $3.75\div\frac{3}{4}\rightarrow3.75\div0.75\rightarrow5$ miles/hour Joyce walks one-tenth of a mile farther in 1 hour, or 528 ft. (One mile is 5,280 ft.)
115.	Use rate x time = distance Train: rate = 60; time = t Plane: rate = 310; time = t - 2 Use either option, the distance is the same, so you can set up an equation. $d_{train}=d_{plane}\rightarrow R_{train}T_{train}=R_{plane}T_{plane}$ $60t=310(t-2)\rightarrow t=2.48$ hours The distance between the cities is $60(2.48)$ or 148.8 miles.
116.	average speed = $\frac{\text{total distance}}{\text{total time}}$ total distance: 10 + 20 + 32 = 62 miles total time: 60 minutes = 1 hour average speed: 62 miles per hour
117.	average speed = $\frac{\text{total distance}}{\text{total time}}$ Note: If you run 12 mph for 20 minutes, you will run a distance of 4 miles. total distance: 1 + 4 = 5 miles total time: 30 minutes = 0.5 hour average speed: 10 miles per hour
118.	a. The $500 account increases by 4%, or $20. The other account increases by $41 - $20, or $21. If the 3% account increases by more money ($21) that the 4% account ($20), but it has a lower interest rate, you must have put more money in the second account. b. If a 3% increase equals an increase of $21, you can solve an equation that asks, "3% of what number equals $21?" $\rightarrow0.03x=21\rightarrow x=700$
119.	You put the same amount of money in the accounts, or $2,000 each. The 2% account increases by 2% of $2,000, or

	$40. The variable interest rate account increases by $60. You can solve the equation $2000x=60\rightarrow x=0.03$. The variable account increased by 3%.
120.	Create a system of equations. Equation 1: $A + B = 1{,}000{,}000$ Equation 2: $1.045A + 1.06B = 1{,}056{,}250$ A = $250,000 B = $750,000
121.	The circle has a larger area. area of square: 7x7 or 49 cm^2 area of circle: $\pi(4)^2\rightarrow16\pi\rightarrow50.3\ cm^2$
122.	The area increases by 300%. If the sides double, the area is 10x10 or 100 square inches, which is 4 times larger.
123.	x = 2, -1 $\frac{-x+2}{x(x-1)}\cdot\frac{2}{2}=\frac{x}{x-1}\cdot\frac{2x}{2x}-\frac{4}{2(x-1)}\cdot\frac{x}{x}$ $\frac{-2x+4}{2x(x-1)}=\frac{2x^2}{2x(x-1)}-\frac{4x}{2x(x-1)}$ $-2x+4=2x^2-4x\rightarrow2x^2-2x-4=0$ $2(x^2-x-2)=0\rightarrow2(x-2)(x+1)=0$
124.	$\frac{5x+6}{x(x+2)}$ $\frac{6}{(x+2)(x-2)}\cdot\frac{x-2}{3}+\frac{3}{x}\rightarrow\frac{2}{x+2}+\frac{3}{x}$ $\rightarrow\frac{2\cdot x}{(x+2)\cdot x}+\frac{3(x+2)}{x(x+2)}\rightarrow\frac{2x+3x+6}{x(x+2)}$
125.	a. $x<4$ 4 b. $-2x\le12\rightarrow x\ge-6$ -6
126.	$-8y<-2x+40\rightarrow y>\frac{1}{4}x-5$
127.	a. 2 - 3(-2) = 8 b. 4 c. x = -4 or 3 d. $f[g(1)]\rightarrow f[-3]\rightarrow2-3(-3)\rightarrow11$

© Alex Joujan, 2020

128.	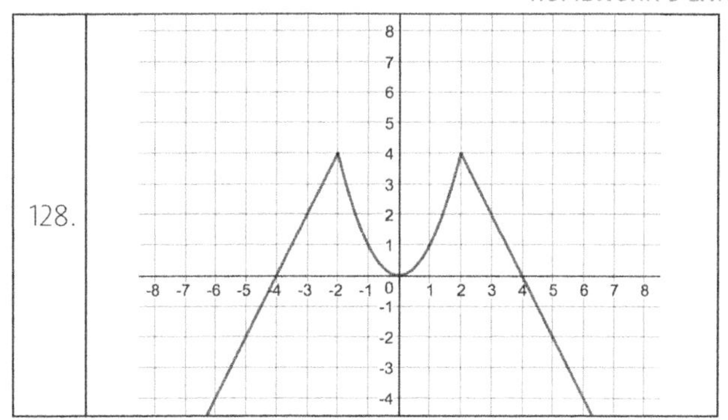

 © Alex Joujan, 2020

 © Alex Joujan, 2020

Made in the USA
Monee, IL
18 July 2024

62031056R00059